# Detection of Karst Voids at Deep Pile Foundation

Liu Liu · Zhenming Shi · Ming Peng · Shaojun Li ·
Fengjuan Tao

# Detection of Karst Voids at Deep Pile Foundation

同濟大學出版社
TONGJI UNIVERSITY PRESS

Springer

Liu Liu
Institute of Rock and Soil Mechanics
Chinese Academy of Sciences
Wuhan, China

Zhenming Shi
College of Civil Engineering
Tongji University
Shanghai, China

Ming Peng
College of Civil Engineering
Tongji University
Shanghai, China

Shaojun Li
Institute of Rock and Soil Mechanics
Chinese Academy of Sciences
Wuhan, China

Fengjuan Tao
Wuhan Changsheng Engineering
Exploration Technology Development Co., Ltd.
Wuhan, China

ISBN 978-981-97-5833-3 ISBN 978-981-97-5834-0 (eBook)
https://doi.org/10.1007/978-981-97-5834-0

Jointly published with Tongji University Press Co., Ltd.
The print edition is not for sale in China (Mainland). Customers from China (Mainland) please order the print book from: Tongji University Press Co., Ltd.



This Springer imprint is published by the registered company Springer Nature Singapore Pte Ltd.
The registered company address is: 152 Beach Road, #21-01/04 Gateway East, Singapore 189721, Singapore

# Contents

# Chapter 1
# Introduction

## 1.1 Prologue

Karst voids and sinkholes are widely and randomly developed near the subsurface (Ford and Williams 2007). Taking China as an example, China is one of the countries with the most extensive development of karst geology in the world, with a total area of 3.443 million square kilometers, accounting for 35.86% of the country's land area. The irregular distribution and filling of karst, complex karst water conditions, and fluctuating soil-rock interfaces in complex karst geological environments pose significant challenges to the site selection, design and construction of large-scale engineering projects. Engaging in engineering construction without a clear understanding of the development of karst in the geological strata is extremely dangerous and can lead to serious land subsidence, collapse, and even widespread impacts on the safety of surrounding structures. In order to meet the urgent requirements of regional economic development and connectivity, major infrastructure projects such as highways and railways are rapidly developing and will inevitably cover vast areas of karst development. In fact, frequent engineering-induced karst geological disasters have occurred in provinces such as Guangxi, Guizhou, Guangdong, Yunnan, Tibet, Hunan, Hubei, Shandong, Shanxi, and Sichuan in China, resulting in significant loss of life and property. These disasters have become a bottleneck restricting the planning of China's highway and railway networks and urbanization construction, and will continue to affect the long-term operation of existing projects.

Pile foundations are one of the most common technologies used to support heavy buildings in karst areas (Zhou and Beck et al. 2011). The inspection of the karst void distribution at the pile location is the top priority in the design and construction of pile foundations. The design of the bearing capacity of the pile foundation mainly includes pile end resistance and pile side friction (Peng et al. 2021). Karst voids beneath the bottom of the pile hole significantly affect the pile end resistance. The bearing capacity of the rock socketed pile is closely related to the thickness of the roof of karst voids. The standard (TNSCG 2009) indicates that it is necessary to ensure



L. Liu et al., *Detection of Karst Voids at Deep Pile Foundation*,
https://doi.org/10.1007/978-981-97-5834-0_1

that there is no karst void within the range of 3 times pile diameters or at least 5 m beneath the bottom of the pile. On the other hand, the pile side friction can be divided into two parts: the side friction in the overlying soil layer and the pile side friction in the bedrock. The karst voids on the side of the piles likely result in the loss of the surrounding soil and rock, which leads to a decrease in the pile side friction. Cast-in-place piles are increasingly used in karst area engineering due to the advantages of no limitation of stratum changes, no need for pile extension and pile cutting, low noise, and less ground settlement caused by pumping. However, the developments of karst voids at pile locations bring extra difficulties to the construction of pile foundations, which can easily cause pile hole collapse, deviation, sticking, and slurry leakage.

In authors' point of view, inspection of karst voids at pile locations faces the following three problems: (1) The inspection should be construction friendly. The detection should not interrupt the construction process since numerous piles are drilled in a single project. The existing boreholes and pile holes can be used for precise detection during construction. (2) The multi-solution problem of detection data interpretation needs to be mitigated. The inspection environment, such as voids, rock fracture, slurry, and pile holes, makes the wavefield complex. It is necessary to propose an effective signal processing method to improve the signal-to-noise ratio and extract effective void reflections. (3) The signal analysis method should be easily understandable. The inspection device and signal analysis method are used if the operators have geophysical or civil engineering backgrounds.

## 1.2 Overview of Detection Method for Karst Voids at Pile Location

The inspection technologies of karst voids in the shallow subsurface can be divided into surface-based and borehole-based detection.

(1) surface-based detection

The high-density resistivity method is based on the conductivity differences between underground media. It determines the geologic bodies with different resistivities by measuring the distribution of electrical currents in different formations under the influence of an external electric field. With the advancement of electronic technology, high-density resistivity instruments focused on improving acquisition efficiency and resolution have rapidly developed. The high-density resistivity method has the ability to visually and vividly reflect the morphology and characteristics of electrical anomalies in cross sections (Goodman et al. 2005). Distributed, high-precision, and multi-channel configurations are the development trends in modern geological exploration. The new distributed high-density resistivity instrument, with switching and acquisition modules distributed among various electrodes, is expected to become the mainstream in the future. Practical and theoretical research results have shown that this exploration method yields high-quality data for voids detection in areas with gentle terrain and good grounding conditions. After inversion, the results can effectively reflect the actual situation. However, in areas with poor grounding conditions, the inversion results are less accurate, and structures such as near-surface concrete or asphalt roads and various pipelines tend to be amplified, leading to misinterpretation. Additionally, the lateral and vertical resolutions of the high-density resistivity method are closely related to electrode arrangement and spacing selection. When measuring potential differences, anomalies in the vertical direction can affect the measurements below, resulting in higher lateral resolution than vertical resolution for the high-density resistivity method (Liu et al. 2017). This poses challenges in detecting deep karst formations, especially when surface detection is difficult.

The shallow seismic method is based on the impedance differences between underground media and utilizes information such as the intensity, spectrum, phase, wavelength, and travel time of reflected waves to detect voids at pile locations (Li et al. 2017). As a large-scale geophysical method for engineering exploration, the shallow seismic reflection method serves as a preliminary means of geo-investigation. The shallow seismic method is non-destructive, efficient, and accurate, making it well-suited for detecting shallow underground spaces. The effectiveness of the shallow seismic method is often dependent on factors such as shot spacing, source frequency, and bandwidth (Li et al. 2013). However, as a surface geophysical method, it has limited investigation depth, especially when there is a thick overlying soil layer or deep pile foundations, making it difficult to accurately detect voids at the bottom of the pile. It is more suitable for regional classification and provides limited geological information for pile foundation construction and design.

Xia et al. (1999, 2002a, 2002b, 2005, 2012) have conducted a systematic study on determining shallow surface wave velocities using high-frequency multi-channel surface wave methods since 1999. This high-frequency surface wave technology has received great attention in the field of shallow surface geophysics. In traditional seismic exploration on a large scale, surface waves and body waves can be well separated, and surface waves are usually removed before the inversion process. However, at the small scale of shallow surfaces, it is difficult to separate body waves from surface waves, especially when dealing with field data in shallow seismic exploration. Surface waves typically account for two-thirds of the total seismic energy. Due to the dispersion of surface waves in heterogeneous media (Thomson 1950), they can be analyzed to obtain model parameters, typically representing near-surface shear wave velocity models. The high-frequency surface wave method can be used for detecting shallow buried voids, but it is challenging to accurately identify voids in thick overlying soil layers or in bead-like karst formations. The test profiles provide 2D shear wave velocity profiles, which make it difficult to accurately determine the development of karst in pile locations. Overall, with the increasing application of ultra-long piles, traditional surface geophysical methods are no longer able to meet the requirements for detecting karst at pile locations. Borehole geophysical methods, with advantages such as less interference and closer proximity to the target, should be more suitable for karst detection around boreholes.

(2) borehole-based detection

Cross-hole elastic wave computed tomography (CT) utilizes elastic wave rays to penetrate geological formations. By observing and inverting the travel times and attenuation of seismic waves, the internal structure of the geological formation can be imaged. The typical cross-hole spacing is around 10 m. CT images provide a visual representation of the distribution boundaries and the degree of fragmentation and distribution of various rock and soil formations (Duan et al. 2017). The resolution of elastic wave CT mainly depends on the observation geometry, ray density, source and receiver spacing, background wave velocity, and measurement errors. Due to the fact that elastic waves propagate along the fastest travel path rather than in a straight line, neither the straight-line algorithm nor the curved ray algorithm can accurately determine the true wave velocity of the geological formation. Moreover, elastic wave CT images are unable to observe velocity anomalies lower than 85 to 64% of the background velocity. The spacing between the two boreholes in elastic wave CT testing is generally within the range of 5–30 m. Various seismic sources have been used, including explosive sources, spark sources, and magnetostrictive sources. Underground seismic sources commonly use pulsed spark sources, which have the advantages of continuous excitation, minimal energy loss, deep penetration, low delay, and good consistency. Electrode discharge is used to avoid the degradation of excitation performance caused by corrosion of exposed electrodes. In terms of receivers, hydrophones with wide response bandwidth, high sampling accuracy, high sensitivity, and low nonlinear distortion are typically used as receiver arrays. Alternatively, downhole receiver arrays with piezoelectric crystals and independent amplifiers can also be used for signal reception. In elastic wave CT detection, boreholes need to be drilled on both sides of the pile location, which requires a significant amount of drilling work. To prevent hole collapse during testing, PVC casings must be inserted to the bottom of the hole after drilling to the design depth. In practice, due to the filling of soft clay or sand in voids, drilling sediments often result in insufficient effective detection depth. To ensure that PVC can pass through the voids, steel casings need to follow till the bottom of the hole during drilling. After inserting the PVC pipe to the bottom, the steel casing needs to be removed, which requires significant effort and cost.

Borehole radar technology utilizes the propagation characteristics of electromagnetic waves in geological formations to obtain subsurface information (Tsoflias et al. 2015; Tsoflias 2014). Single-borehole radar distinguishes the lithology and geometric features of target bodies based on the travel time, amplitude, and phase of the reflected waves from subsurface anomalies. Borehole radar technology overcomes the limitations of conventional ground-based radar and geological exploration techniques, providing a more extensive and intuitive range of information about subsurface targets. Its main application areas include environmental assessment, engineering geology, geotechnical engineering, archeology, hydrogeology, and site selection for high-level radioactive waste storage facilities. Zhong et al. (2011, 2012) conducted forward modeling studies on the reflection characteristics of point-like adverse geological conditions, surrounding rock morphology, and filling effects using borehole radar, verifying the feasibility of using borehole radar for detecting adverse geological formations. In terms of equipment, traditional borehole radar systems have omnidirectional transmitting and receiving antennas, which can only provide information about the depth and distance of the target geological bodies from the borehole, but cannot determine their orientation.

Note that a significant amount of research has been conducted in shallow subsurface geophysical exploration, making important contributions to the investigation of adverse geological conditions in the shallow subsurface. However, most detection techniques still have issues in terms of their adaptability to engineering construction, lacking high-precision, large-scale, and deep geophysical exploration methods that are closely integrated with engineering projects. The existing techniques used in engineering are far behind the speed and scale of large-scale infrastructure construction in China, making it difficult to ensure rapid and accurate detection of the geological conditions of the bearing layers for building foundation engineering. The "Code for Geotechnical Investigation of Foundation Engineering" (TNSCG 2009) specifies that there should be no weak geological formations within three times the diameter of the pile bottom or within a five-meter range.

Currently, drilling is irreplaceable in engineering due to its visual nature and reliability. Compared to surface geophysical methods, drilling geophysical exploration has clear advantages, such as minimal interference and proximity to the target. Since at least one borehole is required for every pile location in karst regions, single-hole detection technology shows strong adaptability to engineering. Time–frequency analysis and full waveform inversion have been widely used in traditional seismic exploration, but there has been little research on acoustic wave detection in boreholes. However, the research experience gained from seismic data analysis and inversion in shallow subsurface seismic data is helpful for analyzing borehole seismic data. Therefore, the development of a geophysical exploration system that combines drilling and pile holes, and the analysis of the detection using time–frequency analysis and full waveform inversion, is an important approach for fine detection of karst at pile locations.

## 1.3 Book Outline

This book introduces a state-of-the-art and newly developed detection system, comprising of specialized equipment, algorithms, and software, designed to identify karst voids during deep cast-in-place pile foundation construction. The system includes a multi-frequency borehole sonic probe, a single borehole radar, two pile hole sonar probes, and corresponding signal analysis method. Several field and synthetic tests have been conducted to detect karst voids during cast-in-place pile foundation construction, and the results demonstrate that this technology can offer detailed geological information for pile design and ensure the safety of foundation construction. Chapter 2 introduces the overall approach for karst detection at pile locations. Chapters 3 to 7 focus on the detection of karst voids around boreholes, with Chapters 7 to 9 specifically addressing the detection of voids beneath the bottom of pile holes. The specific contents of each chapter are as follows:

Chapter 2—Overall strategy: This chapter analyzes in detail the construction process of drilled shafts and proposes specific timing and detection strategy based on the construction process.

Chapter 3—Borehole sonic detection equipment: This chapter presents the overall framework design of the borehole sonic detection system, including the design of the detection host, the structure of the downhole tool, the configuration of the high-frequency piezoelectric acoustic wave transmitter, the low-frequency magnetostrictive transmitter, and the wideband directional receiver. The detection process is also designed.

Chapter 4—Borehole sonic detection analysis method: This chapter uses 2D finite-difference numerical simulation to study the process of borehole sonic detection of karst voids. The detection principles are revealed through the analysis of typical wavefields and wave profiles. The chapter also proposes signal processing methods for borehole sonic detection of karst voids, validates the detection hardware and signal analysis methods through controlled laboratory experiments, and a bridge field case.

Chapter 5—Inversion of borehole sonic detection: This chapter investigates full waveform inversion methods applicable to near-zero-offset data from borehole sonic detection. It proposes target functions and gradient computation methods suitable for zero-offset spacing in boreholes. The effectiveness of full waveform inversion methods is analyzed using numerical simulation data. Field borehole tests are conducted to perform full waveform inversion on field data, and the results are validated using adjacent boreholes and cross-hole tomography.

Chapter 6—Borehole radar detection method: This chapter presents the detection equipment and analysis methods for single-hole borehole radar as an effective complement to borehole sonic detection. The feasibility of borehole radar detection is analyzed using measured data from karst regions.

Chapter 7—Pile hole sonar detection equipment: This chapter presents the overall framework design of the sonar detection system for voids at the bottom of pile holes, including the design of the detection host, the structure of the probe in the pile hole, and the configuration of the one-transmitters-four-receivers probe. The detection process is also designed.

Chapter 8—Pile hole sonar detection analysis method: This chapter calibrates numerical simulation parameters based on field detection data. Then we uses 3D finite-difference numerical simulation to analyze the sonar detection process for voids beneath the bottom of pile holes. The detection principles are revealed through the analysis of wavefields, hypothetical detection wave profiles, and specific detection signals. Qualitative signal analysis methods and time–frequency analysis methods based on generalized S-transform are proposed based on the detection principles. The qualitative analysis methods and time–frequency analysis methods are validated using numerical simulation signals and field case studies.

Chapter 9—Advanced Pile hole sonar detection system: This chapter proposes a one-transmitters-two-receivers form of pile hole detection, analyzes the essential differences between interference noise and useful signals, and theoretically proposes an index called instantaneous phase difference intensity (IPDI) for automatically extracting effective reflected waves from voids. We also define the calculation method of IPDI. Sensitivity analysis of the defined index and calculation methods is performed using numerical simulation signals, and the instantaneous phase difference intensity is calculated in multiple field cases. The detection results are validated using borehole television and cross-hole tomography.

Chapter 10—Complete application: This chapter provides a detailed investigation and construction survey case study where borehole sonic detection and pile hole sonar detection are both applied to the same project. The effectiveness of the entire detection system is validated using numerical simulation and field data.

Chapter Back matter—Conclusion and further research suggestion: This chapter systematically summarizes the proposed pile location karst detection system and proposes future research suggestions.

## References


Duan C, Yan C, Xu B et al (2017) Crosshole seismic CT data field experiments and interpretation for karst caves in deep foundations[J]. Eng Geol 228:180–196

Ford D, Williams PD (2007) Karst hydrogeology and geomorphology. John Wiley & Sons

Goodman N, Leuschen C, Plumb R et al (2005) Imaging subsurface cavities using geoelectric tomography and ground-penetrating radar[J]. J Cave Karst Stud Nat Speleolog Soc Bull 67(3):174–181

Li S, Zhou Z, Li L et al (2013) Risk assessment of water inrush in karst tunnels based on attribute synthetic evaluation system[J]. Tunn Undergr Space Technol 38:50–58

Li S, Liu B, Xu X et al (2017) An overview of ahead geological prospecting in tunneling[J]. Tunn Undergr Space Technol 63:69–94

Liu B, Liu Z, Li S et al (2017) An improved time-lapse resistivity tomography to monitor and estimate the impact on the groundwater system induced by tunnel excavation[J]. Tunn Undergr Space Technol 66:107–120

Peng M, Wang D, Liu L et al (2021) Near-surface geological structure seismic wave imaging using the minimum variance spatial smoothing beamforming method. Appl Sci 11(22):10827

The National Standards Compilation Group of People's Republic of China (2009) GB50021—2009 code for investigation of geotechnical engineering. China Architecture and Building Press, Beijing, pp 52–55

The National Standards Compilation Group of People's Republic of China. GB50021 (2009) Code for investigation of geotechnical engineering[S]. China Architecture and Building Press, Beijing, pp 52–55.(in Chinese)

Thomson WT (1950) Transmission of elastic waves through a stratified solid medium[J]. J Appl Phys 21(2):89–93

Tsoflias GP, Perll C, Baker M et al (2015) Cross-polarized GPR imaging of fracture flow channeling[J]. J Earth Sci 26(6):776–784

Tsoflias GP, Borehole GPR (2014) Direct and refracted arrivals for determination of subsurface void dimensions[C]. In: Near surface geoscience 2014–20th European meeting of environmental and engineering geophysics. European association of geoscientists & engineers, pp 1–5

Xia J, Miller RD, Park CB (1999) Estimation of near-surface shear-wave velocity by inversion of Rayleigh waves[J]. Geophysics 64(3):691–700

Xia J, Miller RD, Park CB et al (2002a) A pitfall in shallow shear-wave refraction surveying[J]. J Appl Geophys 51(1):1–9

Xia J, Miller RD, Park CB et al (2002b) Comparing shear-wave velocity profiles inverted from multichannel surface wave with borehole measurements[J]. Soil Dyn Earthq Eng 22(3):181–190

Xia J, Chen C, Tian G et al (2005) Resolution of high-frequency Rayleigh-wave data[J]. J Environ Eng Geophys 10(2):99–110

Xia J, Xu Y, Luo Y et al (2012) Advantages of using multichannel analysis of Love waves (MALW) to estimate near-surface shear-wave velocity[J]. Surv Geophys 33(5):841–860

Zhong S, Wang C, Wu L et al (2011) Borehole radar response characteristics of point unfavorable geo-bodies: forward simulation on its geometric effect[J]. Rock and Soil Mech 32(5):1583–1588

Zhong S, Wang CY, Wu LX et al (2012) Borehole radar response characteristics of point unfavorable geo-bodies: forward simulation of its surrounding rock and filling condition[J]. Rock Soil Mech 33(4):1192

Zhou W, Beck BF (2011) Engineering issues on karst management. Springer, Dordrecht, pp 9–45

# Chapter 2
# Overall Strategy

## 2.1 Detection Strategy

To design an appropriate detection strategy and device for cast-in-place pile construction in karst areas, it is essential to review the construction process. This process can be divided into four distinct phases, as illustrated in Fig. 2.1. The following are the key phases:

Advanced Borehole Investigation (Fig. 2.1a): Conducting comprehensive borehole investigations at the center of the piles is imperative in areas with complex karst voids. These investigations provide valuable geological information that helps determine the drilling parameters for the pile holes.

Slurry and Steel Case Protection (Fig. 2.1b): During pile hole drilling, the use of slurry and steel casing is necessary. The steel case should be positioned at the bottom of the lowest karst void encountered by the pile hole. This protective measure ensures the stability and integrity of the pile during the construction process.

Precast Steel Cage Installation (Fig. 2.1c): After verifying the quality of the pile hole and the thickness of sediment at the pile bottom, a precast steel cage is installed into the pile hole. This cage provides structural reinforcement to the pile.

Concrete Filling (Fig. 2.1d): The pile hole is then filled with concrete to complete the construction process. It is crucial to note that in the illustrated scenario, the advanced borehole investigation successfully identified the middle half-filled karst void (Fig. 2.1a), while the other two karst voids were not detected throughout the entire construction process. However, both the upper and lower karst voids can significantly impact the pile's side friction and end-bearing capacity, posing high-risk factors for building stability. Therefore, a more detailed investigation is urgently required to accurately assess the geological conditions and provide correct information for construction planning.

Based on the discussed construction process, a comprehensive investigation strategy has been devised to identify hidden karst voids at pile locations. The strategy is outlined in Fig. 2.2. It involves the following steps: Borehole investigation is an

 
L. Liu et al., *Detection of Karst Voids at Deep Pile Foundation*,
https://doi.org/10.1007/978-981-97-5834-0_2

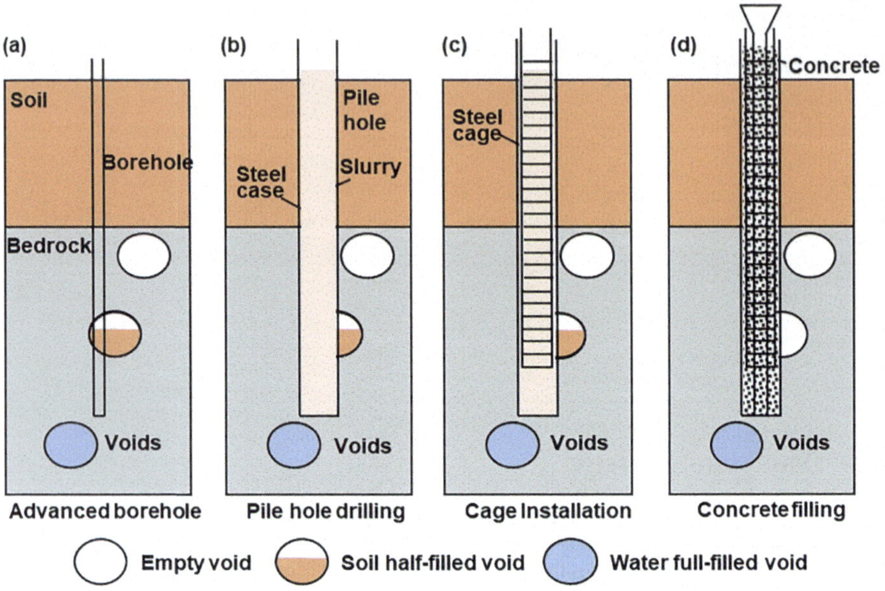


**Fig. 2.1** Four phases (**a**–**d**) of cast-in-place pile construction

indispensable part of the overall strategy. Borehole logs and cores provide direct information about the underground geological conditions, despite the smaller size of the borehole compared to the designed pile (Fig. 2.2a). Downhole tools can be utilized to reach the bottom of the borehole and investigate voids near the borehole. Sonic tests with multiple frequencies, such as 2 and 40 kHz, can be conducted prior to drilling the pile hole. The high-frequency sonic test captures images of the borehole wall and the immediate vicinity (within 1 m) around the borehole (red area in Fig. 2.2b). The low-frequency sonic test can detect a wider range (within 3 to 6 m), as depicted by the blue area in Fig. 2.2c. Table 2.1 provides detailed parameters for the borehole detection. The process of drilling the hole for the drilled shaft presents another crucial opportunity to identify geological anomalies through geophysical detection in the pile hole. A geophysical probe, as shown in Fig. 2.2d, is used to reach the bottom of the pile hole and transmit/receive sonar waves in a liquid environment with slurry coupling. Sonar detection in the pile hole aims to identify anomalies beneath the pile hole bottom, providing feedback for the drilling process and enabling the assessment of the final end-bearing rock quality. Sonar detection can be performed during breaks in the pile hole drilling process (Fig. 2.2d) or once the drilling is completed (Fig. 2.2e). This pile hole-based sonic detection enhances the geological information around and beneath the piles. By combining the borehole and pile hole-based sonic detection, the overall investigation strategy provides a comprehensive approach to obtain detailed geological information for pile construction projects.

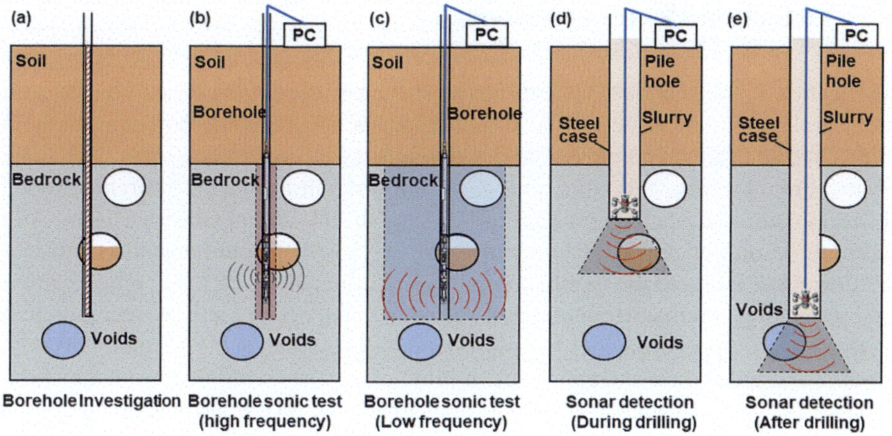


**Fig. 2.2** Scheme of detection of karst voids during cast-in-place pile construction

**Table 2.1** Parameters of the downhole probes

| Probe type | Borehole multi-frequency sonic probe | | 1st Pile hole sonar probe | 2nd Pile hole sonar probe |
|---|---|---|---|---|
| Construction period | Borehole investigation | | Pile hole drilling | Pile hole drilling |
| Estimated detection range | HM*: within 1 m around boreholes | LM*: within 3 ~ 6 m around boreholes | Within 5 m beneath pile hole | Within 8 m beneath pile hole |
| Detection principle | HM: sonic wave attenuation | LM: Reflected wave | Reflected wave identification based on time–frequency analysis | Reflected wave identification based on instantaneous phase difference analysis |
| Detection environment | Water-filled Borehole, PCV-cased if necessary | | Pile hole slurry | Pile hole slurry |
| Source frequency range | HM: 1–50 kHz | LM: 0.05–7 kHz | 0.2–8 kHz | 0.2–8 kHz |
| Source dominant frequency | HM: 40 kHz | LM: 2 kHz | 3 kHz | 3 kHz |
| Source material | HM: Piezoelectric ceramic | LM: Giant magnetostrictive material | Giant magnetostrictive material | Giant magnetostrictive material |
| Polarity of source | Monopole | Monopole | Monopole | Monopole |
| Receiver material | Piezoelectric ceramic | | Piezoelectric ceramic | Piezoelectric ceramic |

* HM: High-frequency mode, LM: Low-frequency mode
The detection ranges are estimated empirically by the assumed P-wave velocity of weathered limestone and the attenuation features of wave with different frequency

## 2.2 Detection Device Overlook

A detection system of karst void at pile location was developed in accordance with the overall detection strategy. The system consists of a borehole probe, a pile hole probe, and an in situ computer, as depicted in Fig. 2.3 (Liu et al. 2020 and 2021). The in situ computer (Fig. 2.3a) functions as the control unit for both the borehole probe (Fig. 2.3b) and the pile hole probe (Fig. 2.3c and d). It is equipped with a 24-bit A/D sampling resolution and a dynamic range of −7 ~ 130 dB. Communication between the downhole tool and the in situ computer is facilitated through a communication cable, enabling command transmission and data transfer. The depth counter mounted on the ground support provides real-time depth information on the computer display. Figure 2.3c and d illustrate two types of pile hole probes with different acquisition geometries. The first pile hole sonar probe consists of a central transmitted source and four surrounding receivers, while the second sonar probe includes one source and two receivers. The interval between the source and receivers, as well as the spacing between the receivers, is adjusted based on the diameter of the pile hole under investigation.

(1) **Borehole Probe**

The characteristics of the three downhole probes are summarized in Table 2.1. The borehole probe, with a diameter of 49 mm, offers two detection modes utilizing different transmitted frequencies. The borehole sonic probe, depicted in Table 2.1, does not include a casing. The low-frequency source, employing giant magnetostrictive materials, is positioned on the left side. It operates at a dominant frequency of 2 kHz, with a frequency range spanning from 0.05 to 7 kHz. The purpose of the low-frequency borehole sonic investigation is to detect karst voids within a range of

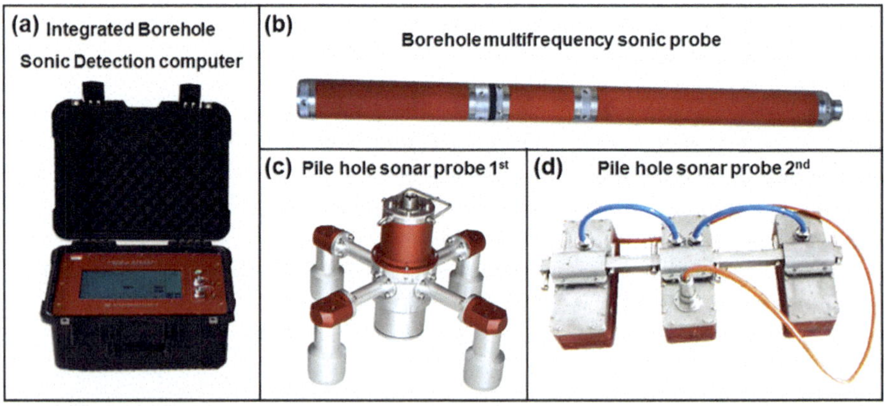


**Fig. 2.3** Detection system of karst voids during cast-in-place pile construction (revised based on Liu et al. 2020 and 2021) **a** detection computer, **b** borehole sonic probe, **c** pile hole sonar probe 1st, **d** pile hole sonar probe 2nd

3 to 6 m surrounding the borehole. This is achieved through the extraction and identification of reflected waves originating from near and behind the voids. The detection range is dependent on the velocity and integrity of the bedrock. On the other hand, the high-frequency ultrasonic source covers a wide frequency bandwidth of 1 to 50 kHz, with a dominant frequency of 40 kHz. This high-frequency test directly assesses the integrity of the borehole wall by analyzing the attenuation of the tube wave (Stoneley wave propagating along the borehole wall). It serves as a reference for the analysis of low-frequency detection data. The high-frequency source is positioned next to the low-frequency source and is constructed using piezoelectric ceramic. A soundproof tube is placed between the sources to prevent wave propagation along the downhole. A receiver with a broad bandwidth is situated to the right of the soundproof tube, allowing for the recording of both low- and high-frequency sonic data.

To ensure the stability of the downhole tool and prevent any issues caused by multiple karst voids, the borehole multi-frequency sonic test requires the use of PVC casing if such voids are detected through borehole logs. The casing helps prevent collapse and sticking of the downhole tool at the voids. Additionally, for effective coupling of sonic waves, the borehole should be filled with water during the test. The detection process involves lowering and raising the downhole tool at intervals of 0.1 m of depth while conducting both low- and high-frequency detection.

(2) **Pile Hole Probe**

The characteristics of the two pile hole sonar probes are provided in Table 2.1, showcasing their distinct acquisition geometries. Both probes employ monopole sources made of Giant magnetostrictive material, enabling the transmission of sonar waves within a frequency range of 0.2–8 kHz in the pile hole slurry. The 1st sonar probe is equipped with four piezoelectric receivers positioned at equal distances from the center source. This arrangement allows for the recording of four azimuthal test signals during each transmission. On the other hand, the 2nd sonar probe features two receivers on one side, providing two recordings with different offsets in a single transmission. Azimuthal detection can be achieved by rotating the sonar probe, facilitated by a three-dimensional compass installed in both probes to measure receiver azimuth during detection.

Due to the requirement for reaching the bottom of the pile hole in the slurry, the pile hole probes equipped with extra weight. During the bored cast-in-place pile construction, these probes operate in the slurry environment of the pile hole. If slurry protection is not implemented, it is recommended to fill the pile hole with a minimum water thickness of 0.2 m. While a single transmission can indicate the presence of voids, accurate identification of void anomalies beneath pile holes necessitates multiple detections conducted at various azimuths.

## 2.3 Analysis Method Overlook

A signal analysis method for detection during cast-in-place pile construction is presented, as illustrated in Fig. 2.4. This method combines borehole and pile hole sonic detection data analyses to evaluate the presence, size, location, and infill of karst voids around the borehole and beneath the pile hole.

During the borehole investigation phase, borehole sonic detection analysis is conducted, encompassing both high- and low-frequency test data analyses. The analysis of high-frequency borehole sonic data focuses on assessing the integrity of rock borehole walls through the calculation of trace signal energy. The amplitude of the signal reflects the quality of the rock, with a larger amplitude indicating better rock quality. Karst voids and rock fractures can be identified by variations in signal energy at different depths. For low-frequency borehole data analysis, a method comprising bandpass filtering, median filtering, and F-K filtering is employed. Bandpass filtering is applied to enhance the signal-to-noise ratio. Median and F-K filters are utilized to extract reflected P-wave events originating from karst voids around the borehole. These filters exploit the slope difference in the wave profile caused by the velocity disparity between the reflected P-wave, converted wave, wave in the borehole, and wave propagating along the borehole wall. By analyzing the common offset wave profile, significant slope differences can be observed. To obtain a detailed image

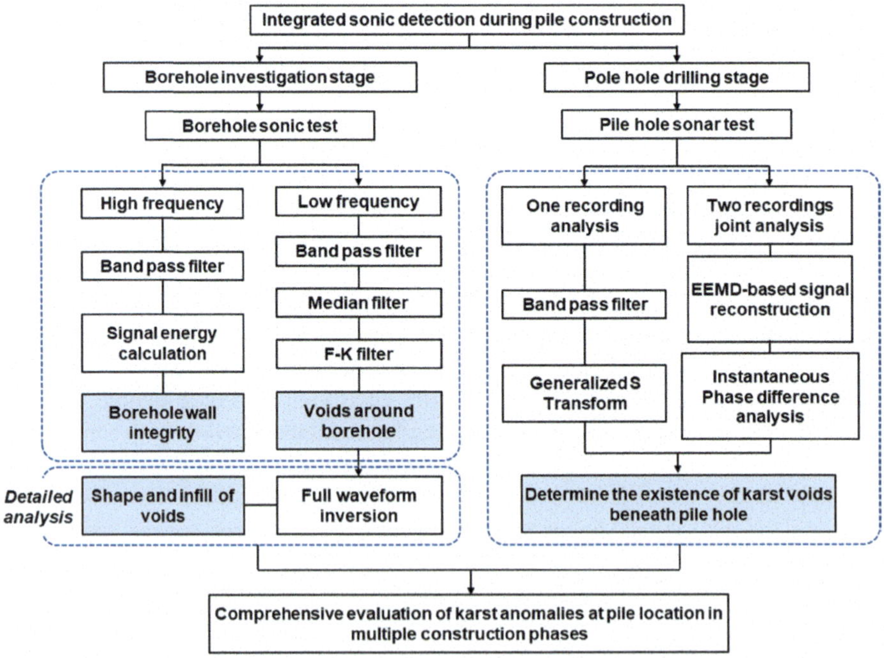


**Fig. 2.4** Signal analysis method of karst voids detection during cast-in-place pile construction

of the velocity field of the rock surrounding the borehole, full-waveform inversion (FWI) is employed for low-frequency borehole data. The inverted velocity model provides insights into the infill and size of the voids.

The analysis of pile hole detection comprises one-recording and two-recording methods, corresponding to the two sonar probes used. In the one-recording method, a generalized S-transform is applied for time–frequency analysis. The presence of voids is indicated by noticeable changes in the time–frequency spectrum, while the initial reflection break is challenging to discern due to background noise interference in the original test signal. For the two-recording method, an instantaneous phase analysis approach is introduced to identify effective reflections from voids by exploiting the phase difference between the reflected P-wave and background noise, primarily composed of multiple surface waves traveling along the pile hole bottom. The analysis method involves ensemble empirical mode decomposition (EEMD)-based signal reconstruction, Hilbert transform, and Integral Instantaneous Phase Difference (IIPD) calculation. Peaks in the IIPD indicate the first break of effective reflections from voids, providing an estimation of their location.

It is important to note that signal energy calculation, bandpass filtering, wave field separation (e.g., median filter and F-K filter), time–frequency spectrum analysis, and IIPD calculation are all processes that aid in estimating the presence of karst voids around the borehole and beneath the pile holes. Detailed information regarding the shape and infill of the voids can be directly identified from the inverted velocity model obtained through FWI. However, the FWI process requires manual selection of inversion parameters and operation.

## References

Liu L, Shi ZM, Peng M et al (2020) A borehole multifrequency acoustic wave system for karst detection near piles. J Appl Geophys 177:104051

Liu B, Wang C, Liu Z et al (2021) Cascade surface and borehole geophysical investigation for water leakage: a case study of the Dehou reservoir, China. Eng Geol 294:106364

Liu L, Shi Z, Tsoflias GP et al. (2021b) Detection of karst cavity beneath cast-in-place pile using the instantaneous phase difference of two receiver recordings. Geophysics 86(1):EN27–EN38

# Chapter 3
# Borehole Sonic Detection Equipment

## 3.1 Prologue

Investigations for the detection of karst cavities can be divided into three stages: preliminary investigation, detailed investigation, and construction investigation (Liu et al. 2017c). The preliminary investigation aims to identify the approximate distribution and development trends of geological phenomena that may affect construction. The detailed investigation focuses on determining the subsurface properties and providing a reliable basis for design and construction. The construction investigation deals with unexpected geological problems encountered during construction. In the preliminary investigation, various surface geophysical methods such as gravimetric prospecting, ground-penetrating radar (GPR) (Zajc et al. 2015; Angelis et al. 2018; Kilic and Eren 2018), multi-channel analysis of surface waves (MASW) (Zeng et al. 2012; Xia et al. 2013; Xia 2014; Hu et al. 2018), electrical resistivity tomography (ERT) (Martínez-Pagán et al. 2013, 2017b; Liu et al. 2017a), or seismic methods (Jaganathan et al. 2010; Chalikakis et al. 2011) can be used to gather information about the general distribution of karst cavities. However, the accuracy of these methods in providing detailed geological information for engineering design may be limited by factors such as the size and depth of the karst cavities and the thickness of the overburden (Laskow et al. 2011; Duan et al. 2017). During the detailed investigation, additional boreholes and borehole geophysical techniques such as cross-hole methods (resistivity, radar, and seismic computed tomography) and single-hole methods (e.g., guided wave method (Li et al. 2005)) are employed to detect karst cavities in or near the pile locations. However, cross-hole methods require multiple boreholes to accurately identify the cavities for each pile, leading to increased time and economic costs. This limits their widespread application. Furthermore, the range of investigation using guided wave detection may be insufficient, particularly for large-diameter piles, and the azimuth of the cavities cannot be determined using single borehole-guided wave imaging (Liu et al. 2017c). In the construction investigation stage, sonar detection is a nondestructive testing method used to ensure the



L. Liu et al., *Detection of Karst Voids at Deep Pile Foundation*,
https://doi.org/10.1007/978-981-97-5834-0_3

quality of the bedrock beneath the pile bottom (Shi et al. 2016 and 2018). However, this method can only be applied during the drilling of the pile.

Advance drilling techniques have become widely used for detecting karst cavities. Typically, at least one borehole is drilled for each pile, but it is challenging to identify all karst cavities using a single borehole due to its limited coverage (TNSCG 2009). To improve the effectiveness of borehole detection, geophysical methods are applied within the borehole. These methods extend the detection range around the pile location and provide precise information about cavity detection for engineering design and construction purposes.

Array sonic logging detection is a commonly employed geophysical method in the oil and gas industry for prospecting and development (Tang et al. 2007; Bing et al. 2011; Wu et al. 2012; Li et al. 2014 and 2019; Wang et al. 2015 and 2017; Hirabayashi et al. 2016 and 2017). However, the existing instrumentation is not suitable for shallower engineering investigations. In civil engineering applications, traditional downhole tools used for array sonic logging are too long to be effectively used. Typically, in pile locations where the bedrock is intact, the piles are socketed only a few meters into the rock. Borehole investigations are conducted to identify suitable bearing layers, typically with a thickness of 3 to 5 m. Consequently, the detection depth range needed is limited to the bottom range of the borehole. However, it is challenging to utilize array sonic logging downhole tools to effectively detect the bottom range of the borehole, particularly when the downhole tool itself exceeds 3 m in length (Che et al. 2014 and 2017; Li et al. 2017 and 2019).

In this chapter, we introduce a detection system called the multi-frequency acoustic wave (MFAW) system for detecting karst cavities near advance drilling boreholes. The system consists of a borehole detector and analysis software. To image the area surrounding the borehole with varying detection ranges and resolutions, a downhole tool is developed with an MFAW transmitter and four azimuthal receivers. The borehole imaging is conducted in two modes: i) High-frequency wave detection is used to identify karst cavities intersecting the borehole and assess the integrity of the rock along the borehole. ii) Low-frequency wave detection is utilized for zero-offset reflection profiling to detect karst cavities located several meters away from the borehole. The azimuth of the cavities is determined by analyzing the signals from the azimuthal receivers.

## 3.2 Multi-frequency Acoustic Wave Detection

To detect karst cavities near the pile location, a borehole MFAW detection system has been developed and implemented, as depicted in Fig. 3.1a. The system utilizes an advance borehole drilled at the center of the pile location, with a smaller diameter than the pile itself (indicated by the dotted line in Fig. 3.1a). The MFAW detection process consists of the following five procedures:

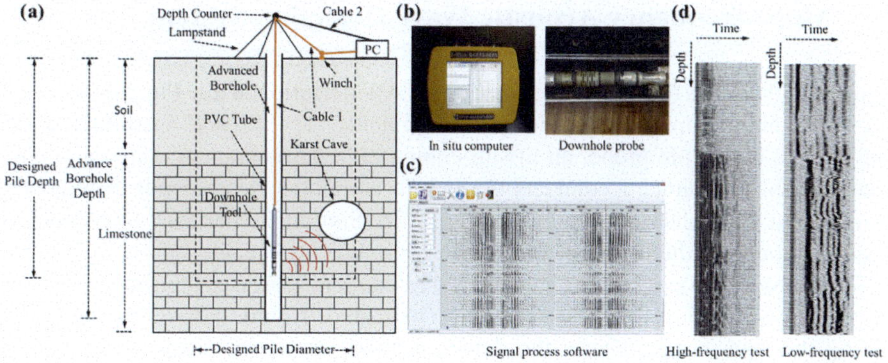


**Fig. 3.1** **a** Schematic diagram of the MFAW detection system deployed in an advance borehole. A karst void is shown within the planned pile outlined by a dashed line. **b** The system controller and downhole probe. **c** Image processing software window. **d** High- and low-frequency data profiles

(1) PVC casing is inserted into the borehole to prevent collapse and shrinkage, as shown in Fig. 3.1. The casing is filled with water to establish acoustic coupling with the bedrock.
(2) A downhole probe (Fig. 3.1b) is carefully lowered into the borehole using ground support. The probe is connected to a controller (in situ computer) via a communication cable (referred to as cable 1 in Fig. 3.1). The depth counter, positioned on the ground support, records the probe's depth and transmits the readings to the controller via cable 2.
(3) The downhole probe is gradually lowered from top to bottom of the borehole at a velocity of 0.1 m/s. At each 0.1 m depth interval, a low-frequency acoustic wave is transmitted. The waves propagate along the borehole, interact with impedance interfaces, and are received and preprocessed by four azimuthal receivers within the probe. The orientation of each receiver is determined using a 3D electric compass. The in situ computer, equipped with signal processing software, analyzes the acoustic profiles along the four azimuths (Fig. 3.1c).
(4) The probe is raised from the bottom to the top of the borehole at a velocity of 0.1 m/s. At each 0.1 m interval, a high-frequency acoustic wave is transmitted and received. Consequently, eight profiles comprising four azimuths of high and low wave frequencies are generated during a logging run (down and up the borehole). The high-frequency data provides information on the immediate vicinity of the borehole, including soil-rock interfaces and fractures. The low-frequency data, on the other hand, allows imaging of geological units farther away from the borehole within the pile region.
(5) After rotating the probe by 30 degrees, steps (3) and (4) are repeated three times. This results in a total of twelve azimuthal test signals, which are comprehensively analyzed.

## 3.3 Multi-frequency Acoustic Wave Detection System

Figure 3.2 illustrates the hardware setup for MFAW detection, while Fig. 3.3 provides a detailed photograph of the relevant components. The MFAW instrumentation comprises a downhole probe, an in situ computer, a depth counter, and a communication cable. Channel 1 connects the in situ computer to the depth counter via communication cable 1, as depicted in Fig. 3.3b. Similarly, channel 2 connects the in situ computer to the downhole probe via communication cable 2.

The detection process can be categorized into two phases: transmitting (indicated by red arrows in Fig. 3.2) and receiving (indicated by blue arrows in Fig. 3.2). During the transmitting phase, the detection mode and parameters, such as shot intervals and transmitter amplitudes, are inputted via the in situ computer. The central processing unit (CPU) receives these parameters and retrieves the real-time downhole depth from the depth counter. When the downhole tool movement aligns with the input shot interval, the CPU sends an electrical signal to trigger the transmitting wave. The transmitting instruction is transmitted through cable 2 to the advanced reduced instruction set computer machines (ARM) located within the downhole

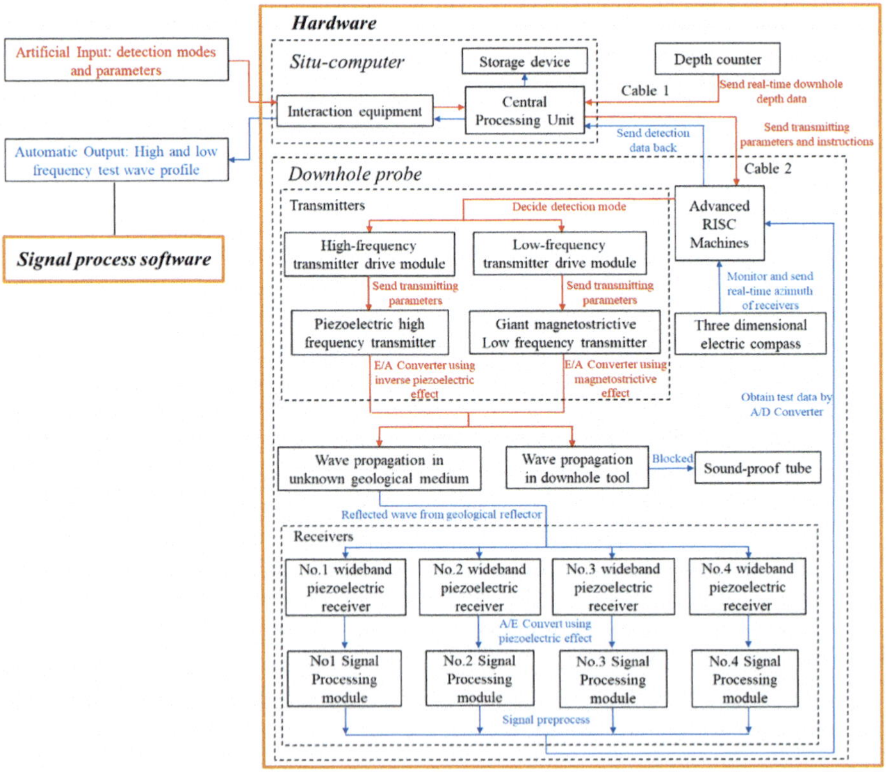


**Fig. 3.2** Schematic of the hardware of the MFAW detection system

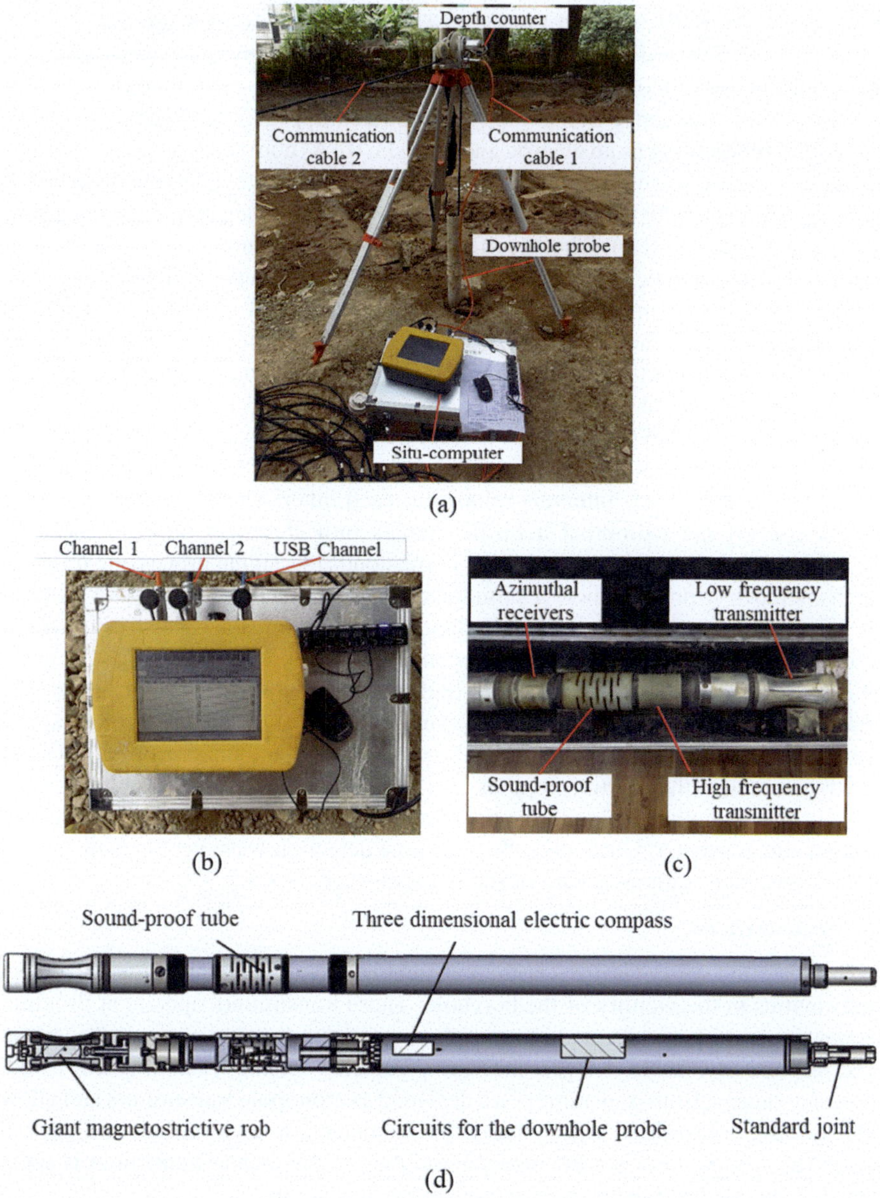


**Fig. 3.3** MFAW detection system: **a** whole detection system, **b** in situ computer, **c** downhole probe, **d** schematic diagram of the downhole probe

probe, as shown in Fig. 3.2 and Fig. 3.3c. The ARM preprocesses the transmitting parameters and forwards them to the corresponding drive module. Depending on whether the probe is being lowered or raised, the appropriate drive module activates the low- or high-frequency transmitter, as illustrated in Fig. 3.3c. The piezoelectric transmitter, functioning as an electric and acoustic (E/A) converter, generates a high-frequency acoustic wave through the inverse piezoelectric effect. On the other hand, the giant magnetostrictive transmitter produces a low-frequency acoustic wave using the magnetostrictive effect. The specifications of the transmitters are described in Sect. 2.1.1. The acoustic wave generated by the transmitter can be divided into two components: the wave propagating outside the downhole tool and the wave traveling along the tool itself. The latter component, considered as noise, is obstructed and attenuated by the sound-proof tube (depicted in Fig. 3.3c and d), ensuring that the receiver does not record these waves.

Wideband azimuthal piezoelectric receivers are utilized to capture both the high- and low-frequency acoustic waves. These receivers consist of four independent sensors positioned at 90° intervals relative to each other. The azimuthal receiving sensors convert the azimuthal acoustic signals into electrical (A/E) signals. In Fig. 3.2, each azimuthal receiving sensor corresponds to a signal processing module, which includes a preamplifier, a numerical control attenuator, an active filter, a programmable amplifier, an analog-to-digital (A/D) converter, and an acquisition circuit. The electrical signals are amplified, filtered, and converted into digital signals within the signal processing module. The digital signal from each azimuthal receiving sensor is transmitted to the ARM and synchronized by the signal synthesis and channel-selection circuits. The recorded azimuthal test signals are combined and then transmitted to the in situ computer via communication cable 1. Simultaneously, the real-time azimuths obtained from the three-dimensional electric compass are sent to the in situ computer. Ultimately, the real-time detection wave profile is displayed on the interactive equipment and stored in the storage device.

(1) **Transmitters**

The detection system utilizes both low- and high-frequency transmitters to detect karst cavities in the vicinity of the borehole. These transmitters operate at different frequencies and are constructed using giant magnetostrictive and piezoelectric materials, respectively. Table 3.1 provides the parameters of the transmitters for both frequency ranges. Both transmitters are designed as monopole transmitters and allow for adjustable transmitted energy based on the detection environment and desired range. The spacing between the receiver and the low-frequency transmitter is set at 0.2 m, while for the high-frequency transmitter, it is 0.1 m.

(a) Giant magnetostrictive transmitter (low frequency)

The low-frequency transmitter is comprised of a giant magnetostrictive rod, a shell with eight vibration strips, and two cover plates, as depicted in Fig. 3.4. The rod is made of giant magnetostrictive material, which consists of a titanium (Ti), dysprosium (Dy), and iron (Fe) alloy compound. The giant magnetostrictive rod undergoes expansion or contraction in response to changes in the applied electric

**Table 3.1** Characteristics of the transmitting transducers

| Transducer type | High-frequency transmitter | | |
|---|---|---|---|
| Polarity | Monopole | Material | Piezoelectric |
| Frequency bandwidth | 1–50 kHz | Working temperature | −10–80 °C |
| Main frequency | 40 kHz | Transmitted energy | 0.1–1 kW<br>Ten adjustable levels |
| Sidelobe | 15 dB | | |
| Transducer type | Low-frequency transmitter | | |
| Polarity | Monopole | Material | Giant magnetostrictive |
| Frequency bandwidth | 0.05–10 kHz | Working temperature | −10–80 °C |
| Main frequency | 3 kHz | Transmitted energy | 5–10 kW<br>Ten adjustable levels |
| Sidelobe | 18 dB | | |

current. The upper and lower cover plates apply a pressure of 8.0 MPa on the giant magnetostrictive rod. Consequently, the vertical contraction and expansion of the rod result in the horizontal vibration of the eight vibration strips, facilitating the transmission of acoustic waves as a monopole source. The shell is filled with epoxy to ensure a waterproof seal.

(b) Piezoelectric transmitter (high frequency)

The high-frequency transmitter is composed of a piezoelectric element, matching and backing layers, and positive and negative electrodes, as depicted in Fig. 3.5. The piezoelectric element is sandwiched between the matching layer and the backing layer. The matching layer improves the acoustic emission efficiency and broadens the frequency range. The backing layer absorbs vibrations and reduces residual oscillations of the transmitter. By utilizing the inverse piezoelectric effect, the electric signal is converted into an acoustic wave. When an electric field is applied to the piezoelectric element in the polarization direction by the electrodes shown in Fig. 3.5, the

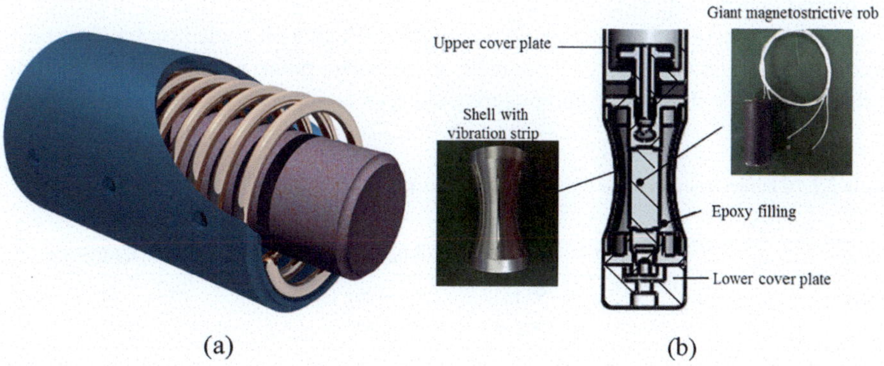


**Fig. 3.4** Schematic diagram of a magnetostrictive transmitter **a** magnetizing coil and ultra-magnetostrictive rod **b** transmitter

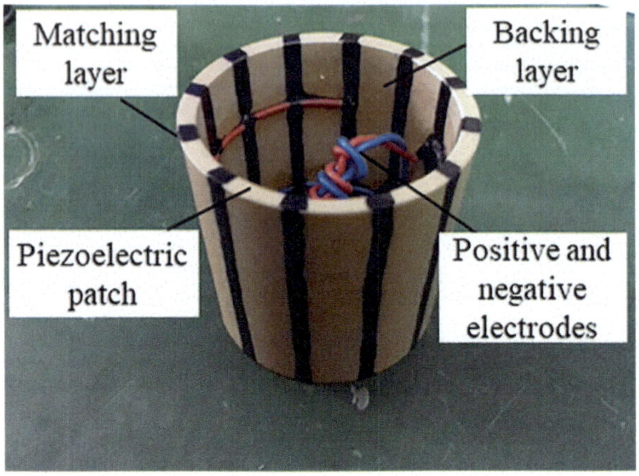


**Fig. 3.5** Schematic of the piezoelectric transmitter

dielectric material undergoes mechanical deformation. A sinusoidal voltage signal is employed to drive the oscillation of the piezoelectric element.

(2) **Azimuthal Receivers**

Wideband azimuthal receivers are utilized to capture both high and low-frequency test signals. The parameters of the azimuthal receivers are presented in Table 3.2. These receivers primarily consist of piezoelectric materials and operate in monopole reception mode, as depicted in Fig. 3.6. Each receiver capsule contains four piezoelectric patches that are filled with silicone oil, enabling efficient coupling between the acoustic wave and the capsule. The four piezoelectric patches are designated as R−x ($x = 1, 2, 3$, and 4), covering a detection range with minimal overlap. Each sensor has a diffusion angle of 120 degrees.

The directional information of the reflected signals can be determined by comparing the energy levels received by the azimuthal sensors. A mandrel is positioned within the four piezoelectric patches, while positive and negative electrodes are placed on opposite sides of the patches. The piezoelectric effect is employed to convert the acoustic wave into an electric signal. When a piezoelectric patch experiences deformation caused by an acoustic stress wave, polarization occurs, resulting in the appearance of positive and negative charges on its surfaces. These charges are

**Table 3.2** Characteristics of the receiving transducers

| Transducer type | Azimuthal receiver | | |
|---|---|---|---|
| Polarity | Monopole | Material | Piezoelectric |
| Frequency bandwidth | 0.05–50 kHz | Working temperature | −10–80 °C |
| Linear frequency | 0.05–20 kHz | Sensitivity | −202 dB |
| Direction diffusion angle | 120° | Pre-amp gain | 0–40 dB |

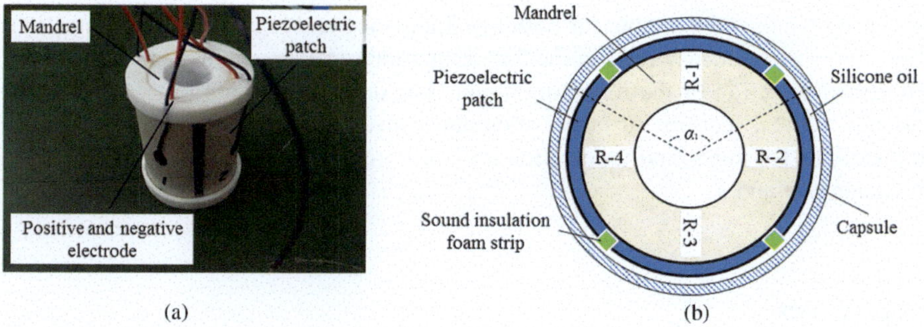


**Fig. 3.6** Schematic of the azimuthal receivers: **a** picture of the receivers without a capsule, **b** schematic of the azimuthal receivers

then detected by the electrodes. Sound insulation material is inserted between adjacent piezoelectric sensors to prevent wave propagation along the mandrel. A three-dimensional electric compass, as shown in Fig. 3.3d, maintains the same azimuth as R-1, allowing for the calculation of the azimuths of the remaining three sensors.

## 3.4 Summary

In this chapter, a borehole multi-frequency acoustic wave detection system is proposed, which includes two different frequency acoustic transmitters, a directional receiver, a depth counter, a three-dimensional electronic compass, and an on-site computer. The borehole multi-frequency acoustic wave detection system collects data point by point along the depth direction of the borehole. The high-frequency acoustic wave is intended to accurately identify karst cavities and fractures encountered by the borehole, as well as to image the rock integrity along the borehole wall. The low-frequency acoustic wave is intended to detect geological anomalies around the borehole related to karst formations. The information on rock integrity obtained through high-frequency acoustic wave detection can assist in interpreting the wave components in the low-frequency detection profiles.

The detection system employs two types of transmitters with different frequencies and principles. The low-frequency transmitter utilizes giant magnetostrictive materials. By changing the current in the coil, the giant magnetostrictive rod undergoes axial elongation and contraction. By applying pressure at both ends and utilizing lateral vibration strips, the vertical vibrations are converted into lateral vibrations, resulting in the emission of acoustic waves with a frequency range of 2–3 kHz in a monopole-like source configuration. The high-frequency transmitter utilizes piezoelectric ceramic materials. Based on the principle of the inverse piezoelectric effect, the piezoelectric ceramic plate vibrates when subjected to polarization, generating acoustic waves with a frequency of 40 kHz. The directional receiver comprises four piezoelectric ceramic plates that receive signals from four different directions. The wideband receiver is suitable for both low-frequency and high-frequency detection. The energy differences between the four azimuthal receivers are used to identify the azimuthal direction of the reflecting target. The detection principles, wave characteristics, typical features of target anomalies, inversion methods, and engineering applications for karst cavity detection around boreholes will be discussed in detail in Chapters 3 and 4.

## References


Angelis D, Tsourlos P, Tsokas G et al (2018) Combined application of GPR and ERT for the assessment of a wall structure at the Heptapyrgion fortress (Thessaloniki, Greece). J Appl Geophys 152:208–220

Bing W, Guo T, Hua W (2011) Extracting near-borehole P and S reflections from array sonic logging data. J Geophys Eng 8(2):308–315

Chalikakis K, Plagnes V, Guerin R et al (2011) Contribution of geophysical methods to karst-system exploration: an overview. Hydrogeol J 19(6):1169

Che X, Qiao W, Ju X et al (2014) An experimental study on azimuthal reception characteristics of acoustic well-logging transducers based on phased-arc arrays. Geophysics 79(3):D197–D204

Che X, Qiao W, Ju X et al (2017) Experimental study on the performance of an azimuthal acoustic receiver sonde for a downhole tool. Geophys Prospect 65(1):158–169

Duan C, Yan C, Xu B et al (2017) Crosshole seismic CT data field experiments and interpretation for karst caves in deep foundations. Eng Geol 228:180–196

Hirabayashi N (2016) Reflector imaging using trial reflector and crosscorrelation: application to fracture imaging for sonic data. Geophysics 81(6):S433–S446

Hirabayashi N, Sakiyama N, Ikegami T (2017) Characteristics of waveforms recorded by azimuthally spaced hydrophones of sonic logging tool for incident plane waves. Geophysics 82:D353–D368

Hu Y, Xia J, Mi B, Cheng F, Shen C (2018) A pitfall of muting and removing bad traces in surface-wave analysis. J Appl Geophys 153:136–142

Jaganathan AP, Allouche E, Simicevic N (2010) Numerical modeling and experimental evaluation of a time domain UWB technique for soil void detection. Tunnel Undergr Space Technol 25(6):652–659

Kilic G, Eren L (2018) Neural network-based inspection of voids and karst conduits in hydro–electric power station tunnels using GPR. J Appl Geophys 151:194–204

Laskow M, Gendler M, Goldberg et al. (2011) Deep confined karst detection, analysis and paleo-hydrology reconstruction at a basin-wide scale using new geophysical interpretation of borehole logs. J Hydrol 406(3–4):158–169

Li X, Guo J, Rao Q (2005) Tube wave detecting method. Chin J Eng Geophys 2(2):129–133

Li J, Tao G, Zhang K et al (2014) An effective data processing flow for the acoustic reflection image logging. Geophys Prospect 62(3):530–539

Li J, Innanen KA, Tao G (2017) Extraction of reflected events from sonic-log waveforms using the Karhunen-Loève transform. Geophysics 82(5):D265–D277

Li J, Innanen KA, Wang B, Tao G (2019) Azimuth ambiguity elimination for borehole imaging using 3D borehole RTM scheme. J Appl Geophys 166:103–111

Liu C, Shi Z, Liu L et al (2017a) State of methods for detecting karst cave on pile location in complicated karst area. Geotech Invest Surv S2:7–13

Liu B, Liu Z, Li S et al (2017b) An improved Time-Lapse resistivity tomography to monitor and estimate the impact on the groundwater system induced by tunnel excavation. Tunnel Undergr Space Technol 66:107–120

Liu B, Liu Z, Li S et al (2017c) Comprehensive surface geophysical investigation of karst caves ahead of the tunnel face: a case study in the Xiaoheyan section of the water supply project from Songhua River, Jilin, Chian. J Appl Geophys 144:37–49

Martınez-Pagán P, Gómez-Ortiz D, Martın-Crespo T et al. (2013) The electrical resistivity tomography method in the detection of shallow mining cavities: a case study on the Victoria Cave. Cartagena (SE Spain): Eng Geol 156:1–10

Shi Z, Liu L, Ming P et al (2016) A sonar detection technology for karst cavities under bored piles and its application. Chin J Rock Mech Eng 35:177–186

Shi Z, Liu L, Ming P, et al. (2018) Sonar detection for karst cavities and fracture under bored pile by generalized S-transform. In: GeoShanghai International Conference, pp 525–533

Tang X, Zheng Y, Patterson D (2007) Processing array acoustic-logging data to image near-borehole geologic structures. Geophysics 72(2):E87–E97

The National Standards Compilation Group of People's Republic of China (2009) GB50021—2009 code for investigation of geotechnical engineering. China Architecture and Building Press, Beijing, pp 52–55

Wang H, Fehler M (2017) The wavefield of acoustic logging in a cased-hole with a single casing–part I: a monopole tool. Geophys J Int 212(1):612–626

Wang H, Tao G (2015) Amethod to determine the strike of interface outside of borehole by monopole borehole acoustic reflections. J Pet Sci Eng 133:304–312

Wu J, Qiao W, Che X (2012) Experimental study on the radiation characteristics of downhole acoustic phased combined arc array transmitter. Geophysics 78(1):D1–D9

Xia J (2014) Estimation of near-surface shear-wave velocities and quality factors using multichannel analysis of surface-wave methods. J Appl Geophys 103:140–151

Xia J, Yin X, Xu Y (2013) Feasibility of determining Q of near-surface materials from Love waves. J Appl Geophys 95:47–52

Zajc M, Celarc B, Gosar A (2015) Structural–geological and karst feature investigations of the limestone–flysch thrust-fault contact using low-frequency ground penetrating radar (Adria–Dinarides thrust zone, SW Slovenia). Environ Earth Sci 73(12):1–13

Zeng C, Xia J, Miller RD, Tsoflias GP, Wang Z (2012) Numerical investigation of MASW applications in presence of surface topography. J Appl Geophys 84:52–60

# Chapter 4
# Borehole Sonic Detection Analysis Method

## 4.1 Prologue

Chapter 3 presents a method for karst cavity acoustic wave detection around boreholes using advanced drilling during the exploration phase. This method utilizes an observation system with near-zero offset, two transmitters with different frequencies, and a directional receiver to detect karst cavities within a certain range around the borehole. However, due to the presence of the borehole, the propagation of acoustic waves can be complex. Especially in complex karst environments, the borehole may intersect multiple cavities, resulting in the superposition of reflected wavefields from multiple cavities. Moreover, the relationship between the two different frequency detection signals, the directivity of the directional receiver, the wavefield characteristics of the quasi-self-receiving detection profile, and the signal processing methods all require in-depth research. This chapter starts from the numerical simulation of acoustic wave detection around boreholes, using two-dimensional time-domain finite difference method to simulate the detection process and wave profiles, revealing the principles of detection. Effective signal processing methods are designed based on the characteristics of the common reflection axis in the wave profiles. The effectiveness of the signal processing methods is verified through numerical simulation signals and laboratory experiments, and the directivity of the receiver is evaluated by comparing the differences in the amplitude of the reflected waves in different directions. Finally, the method and equipment for karst cavity acoustic wave detection around boreholes are applied in the Qingxi Bridge project in Qingyuan, Guangdong province.



L. Liu et al., *Detection of Karst Voids at Deep Pile Foundation*,
https://doi.org/10.1007/978-981-97-5834-0_4

## 4.2 Two-Dimensional Numerical Simulation

### 4.2.1 Numerical Simulation Method

Numerical simulation is crucial for interpreting the detection process, and the staggered grid finite difference (SGFD) method can effectively reduce computational cost and complexity when simulating complex structures. This method has been widely used in various elastic wave detection processes (Jaganathan et al. 2010; Ke and Zheng 2016). Some scholars have successfully simulated small strain piles using the SGFD method with high computational accuracy (Lu et al. 2013; Chang et al. 2016). To simulate the wave propagation process within a limited computational space, absorbing boundaries need to be added outside the computational domain to suppress reflections at the boundaries. The perfectly matched layer (PML) is an effective absorbing boundary that can prevent reflections at both the computational and free boundaries (Berenger 1994; Chew and Liu 1996; Collino and Tsogka 2001). Therefore, the combination of SGFD and PML boundaries is an effective method for simulating acoustic wave detection around boreholes in karst cavities.

In this section, a two-dimensional model of a borehole-rock-cavity system is established. The numerical simulation results are obtained using the two-dimensional viscoelastic wave equation and the staggered grid time-domain finite difference method, allowing for the analysis of the wavefield and wave profiles. The equations, discretization scheme, boundary conditions, initial conditions, and stability conditions for the two-dimensional viscoelastic equation are as follows:

(1) **Two-Dimensional Viscoelastic Equation**

The two-dimensional first-order stress-velocity viscoelastic wave equation is given by Eqs. 4.1–4.5:

$$\frac{\partial v_x}{\partial t} = \frac{1}{\rho}\left(\frac{\partial \tau_{xx}}{\partial x} + \frac{\partial \tau_{xy}}{\partial y}\right) \tag{4.1}$$

$$\frac{\partial v_y}{\partial t} = \frac{1}{\rho}\left(\frac{\partial \tau_{xy}}{\partial x} + \frac{\partial \tau_{yy}}{\partial y}\right) \tag{4.2}$$

$$\frac{\partial \tau_{xx}}{\partial t} = (\lambda + 2G)\frac{\partial v_x}{\partial x} + \lambda\frac{\partial v_y}{\partial y} - \varepsilon_L \tau_{xx} \tag{4.3}$$

$$\frac{\partial \tau_{yy}}{\partial t} = (\lambda + 2G)\frac{\partial v_y}{\partial y} + \lambda\frac{\partial v_x}{\partial x} - \varepsilon_L \tau_{yy} \tag{4.4}$$

$$\frac{\partial \tau_{xy}}{\partial t} = G\left(\frac{\partial v_x}{\partial y} + \frac{\partial v_y}{\partial x}\right) - \varepsilon_T \tau_{xy} \tag{4.5}$$

where $v_x$, $v_y$ are the velocity components at that point; $\tau_{xx}$, $\tau_{xy}$, $\tau_{yy}$ are the stress components at that point; $\rho$ is the density of the medium; $\lambda$ and $G$ are the Lamé constants, which can be calculated using the longitudinal wave velocity $v_p$ and shear wave velocity $v_s$; $\varepsilon_L$ and $\varepsilon_T$ are the damping coefficients for the longitudinal and shear waves, respectively, and they are used to simulate the energy attenuation of waves in the medium (165,166).

(2) **Two-Dimensional Initial Conditions**

At the beginning of the model, all velocity and stress components are zero. A Ricker wavelet is emitted horizontally from the borehole to the surrounding medium. The Ricker wavelet is described by Eq. 4.6.

$$\tau_{xx}(x, y, t) = A \times \left(1 - 2(\pi f_m t)^2\right) \exp\left[-(\pi f_m t)^2\right] \tag{4.6}$$

where (x, y) represents the coordinates of the emission point within the borehole, $t$ is the period of the Ricker wavelet, $f_{\mathrm{m}}$ is the main frequency of the Ricker wavelet, and $A$ is the amplitude of the wavelet. In the numerical simulation of cavities around the borehole, the main frequency of the Ricker wavelet is 3 kHz and the amplitude is 100.

(3) **Higher-order Difference Equation Considering the Two-dimensional PML Boundary**

The two-dimensional viscoelastic wave equation is solved using high-order finite difference in the Cartesian coordinate system. In the numerical simulation of this section, we apply a spatial 10th-order (2N = 10) and second-order in time difference scheme. When the wave propagates to the PML boundary, the PML boundary rapidly attenuates and absorbs the wave energy. In the two-dimensional numerical simulation, as the wave reaches the PML boundary, the wave will decay from two different orthogonal directions. Therefore, different attenuation coefficients $d_x$ and $d_y$ can be used in different directions, as shown in Eqs. 4.7–4.8.

$$d_x = \frac{3}{2}\log\left(\frac{1}{R}\right)V\frac{x^2}{\delta^3}_{\max} \tag{4.7}$$

$$d_y = \frac{3}{2}\log\left(\frac{1}{R}\right)V\frac{y^2}{\delta^3}_{\max} \tag{4.8}$$

where $R$ is the theoretical reflection coefficient, $V$ is the maximum wave velocity, $\delta$ is the thickness of the PML layer, and within the regular computation region, $d_x = d_y = 0$. Based on this, the difference equation for the two-dimensional viscoelastic wave can be obtained. For specific implementation details.

$$v_x = v_x^x + v_x^y \tag{4.9}$$

$$v_{xi,j}^{xt+1/2} = \left(\frac{1}{1+0.5\Delta t \cdot d_x}\right) \cdot \left[(1-0.5\Delta t \cdot d_x)v_{xi,j}^{xt-1/2} + \frac{\Delta t}{\rho \cdot \Delta x}\sum_{n=1}^{N} C_n^{(N)}\left(\tau_{xxi+(2n-1)/2,j}^{t} - \tau_{xxi-(2n-1)/2,j}^{t}\right)\right] \tag{4.10}$$

$$v_{xi,j}^{yt+1/2} = \left(\frac{1}{1+0.5\Delta t \cdot d_y}\right) \cdot \left[(1-0.5\Delta t \cdot d_y)v_{xi,j}^{yt-1/2} + \frac{\Delta t}{\rho \cdot \Delta y}\sum_{n=1}^{N} C_n^{(N)}\left(\tau_{xyi,j+(2n-1)/2}^{t} - \tau_{xyi,j-(2n-1)/2}^{t}\right)\right] \tag{4.11}$$

$$v_y = v_y^x + v_y^y \tag{4.12}$$

$$v_{yi,j}^{yt+1/2} = \left(\frac{1}{1+0.5\Delta t \cdot d_y}\right) \cdot \left[(1-0.5\Delta t \cdot d_y)v_{yi,j}^{yt-1/2} + \frac{\Delta t}{\rho \cdot \Delta y}\sum_{n=1}^{N} C_n^{(N)}\left(\tau_{yyi,j+(2n-1)/2}^{t} - \tau_{yyi,j+(2n-1)/2}^{t}\right)\right] \tag{4.13}$$

$$v_{yi,j}^{xt+1/2} = \left(\frac{1}{1+0.5\Delta t \cdot d_x}\right) \cdot \left[(1-0.5\Delta t \cdot d_x)v_{xi,j}^{yt-1/2} + \frac{\Delta t}{\rho \cdot \Delta x}\sum_{n=1}^{N} C_n^{(N)}\left(\tau_{xyi+(2n-1)/2,j}^{t} - \tau_{xyi-(2n-1)/2,j}^{t}\right)\right] \tag{4.14}$$

$$\tau_{xx} = \tau_{xx}^x + \tau_{xx}^y \tag{4.15}$$

$$\tau_{xxi+1/2,j+1/2}^{xt+1} = \left(\frac{1}{1+0.5\Delta t \cdot d_x}\right) \left[\begin{array}{l}(1-0.5\Delta t \cdot d_x)\tau_{xxi+1/2,j+1/2}^{xt+1/2} \\ +\dfrac{(\lambda+2G)\Delta t}{\Delta x}\displaystyle\sum_{n=1}^{N} C_n^{(N)}\left(v_{xi+1/2+(2n-1)/2,j+1/2}^{t+1/2} - v_{xi+1/2-(2n-1)/2,j+1/2}^{t+1/2}\right)\end{array}\right] - \varepsilon_L \tau_{xxi+1/2,j+1/2}^{xt+1/2} \tag{4.16}$$

$$\tau_{xxi+1/2,j+1/2}^{yt+1} = \left(\frac{1}{1+0.5\Delta t\cdot d_y}\right)$$
$$\left[\begin{array}{l}\left(1-0.5\Delta t\cdot d_y\right)\tau_{xxi+1/2,j+1/2}^{yt+1/2}\\ +\dfrac{\lambda\Delta t}{\Delta y}\sum_{n=1}^{N}C_n^{(N)}\left(v_{yi+1/2,j+1/2+(2n-1)/2}^{t+1/2}-v_{yi+1/2,j+1/2-(2n-1)/2,k+1/2}^{t+1/2}\right)\end{array}\right]$$
$$-\varepsilon_L\tau_{xxi+1/2,j+1/2}^{yt+1/2} \tag{4.17}$$

$$\tau_{xy} = \tau_{xy}^x + \tau_{xy}^y \tag{4.18}$$

$$\tau_{xyi+1/2,j+1/2}^{xt+1} = \left(\frac{1}{1+0.5\Delta t\cdot d_x}\right)$$
$$\left[\begin{array}{l}(1-0.5\Delta t\cdot d_x)\tau_{xyi+1/2,j+1/2}^{xt+1/2}\\ +\dfrac{G\Delta t}{\Delta x}\sum_{n=1}^{N}C_n^{(N)}\left(v_{yi+1/2+(2n-1)/2,j+1/2}^{t+1/2}-v_{yi+1/2-(2n-1)/2,j+1/2}^{t+1/2}\right)\end{array}\right]$$
$$-\varepsilon_T\tau_{xyi+1/2,j+1/2}^{xt+1/2} \tag{4.19}$$

$$\tau_{xyi+1/2,j+1/2}^{yt+1} = \left(\frac{1}{1+0.5\Delta t\cdot d_y}\right)$$
$$\left[\begin{array}{l}\left(1-0.5\Delta t\cdot d_y\right)\tau_{xyi+1/2,j+1/2}^{yt+1/2}\\ +\dfrac{G\Delta t}{\Delta y}\sum_{n=1}^{N}C_n^{(N)}\left(v_{xi+1/2,j+1/2+(2n-1)/2}^{t+1/2}-v_{xi+1/2,j+1/2-(2n-1)/2}^{t+1/2}\right)\end{array}\right]$$
$$-\varepsilon_T\tau_{xyi+1/2,j+1/2}^{yt+1/2} \tag{4.20}$$

$$\tau_{yy} = \tau_{yy}^x + \tau_{yy}^y \tag{4.21}$$

$$\tau_{yyi+1/2,j+1/2}^{xt+1} = \left(\frac{1}{1+0.5\Delta t\cdot d_x}\right)$$
$$\left[\begin{array}{l}(1-0.5\Delta t\cdot d_x)\tau_{yyi+1/2,j+1/2}^{xt+1/2}\\ +\dfrac{\lambda\Delta t}{\Delta x}\sum_{n=1}^{N}C_n^{(N)}\left(v_{xi+1/2+(2n-1)/2,j+1/2}^{t+1/2}-v_{xi+1/2-(2n-1)/2,j+1/2}^{t+1/2}\right)\end{array}\right]$$
$$-\varepsilon_L\tau_{yyi+1/2,j+1/2}^{xt+1/2} \tag{4.22}$$

$$
\begin{aligned}
\tau_{yyi+1/2,j+1/2}^{yt+1} &= \left(\frac{1}{1+0.5\Delta t \cdot d_y}\right) \\
&\left[\begin{array}{l}
\left(1-0.5\Delta t \cdot d_y\right)\tau_{yyi+1/2,j+1/2}^{yt+1/2} \\
+\frac{(\lambda+2G)\Delta t}{\Delta y}\sum_{n=1}^{N} C_n^{(N)}\left(v_{yi+1/2,j+1/2+(2n-1)/2}^{t+1/2} - v_{yi+1/2,j+1/2-(2n-1)/2,k+1/2}^{t+1/2}\right)
\end{array}\right] \\
&- \varepsilon_L \tau_{yyi+1/2,j+1/2}^{yt+1/2}
\end{aligned}
\tag{4.23}
$$

where $\Delta t$ is the time step, and in the numerical simulation of karst cavity detection near the borehole, the time step is 0.002 ms. Cn(N) represents the difference coefficient, and in the case of 10th order difference (2N = 10), $C_1{}^{(5)} = 1.211243$, $C_2{}^{(5)} = -0.08972168$, $C_3{}^{(5)} = 0.001384277$, $C_4{}^{(5)} = -0.00176566$, $C_5{}^{(5)} = 0.0001186795$

(4) **Two-Dimensional Continuity Conditions**

In a two-dimensional space with anisotropic media, appropriate difference treatment is required for the continuity of material parameters $\rho$ and $G$ at the interfaces of different media, as shown in Eqs. 4.24–4.26.

$$
\rho\left(i+\frac{1}{2},j\right) = \frac{\rho(i+1,j)+\rho(i,j)}{2} \tag{4.24}
$$

$$
\rho\left(i,j+\frac{1}{2}\right) = \frac{\rho(i,j+1)+\rho(i,j)}{2} \tag{4.25}
$$

$$
\frac{4}{G_{i+\frac{1}{2},j+\frac{1}{2}}} = \frac{1}{G_{i,j}} + \frac{1}{G_{i+1,j}} + \frac{1}{G_{i,j+1}} + \frac{1}{G_{i+1,j+1}} \tag{4.26}
$$

(5) **Two-Dimensional Stability Conditions**

To ensure the stability of the two-dimensional numerical simulation, the sampling intervals in space and time need to satisfy the following conditions.

$$
v_{\max}\Delta t\sqrt{\left(\frac{1}{\Delta x}\right)^2 + \left(\frac{1}{\Delta y}\right)^2} \le 1 \tag{4.27}
$$

where $\Delta x$ and $\Delta y$ are the spatial sampling intervals, which are set to 0.05 m in this chapter.

### *4.2.2 Numerical Model*

This section calculates three numerical models: the complete formation model, a single cavity model, and the scenario of a cavity encountered during drilling. These models are used to generate wavefield snapshots and seismic profiles, which are then used for signal analysis. The geometric parameters and detection parameters for the numerical models are shown in Table 3.1. The schematic diagrams of the single cavity model (Scenario 2) and the cavity encountered during drilling (Scenario 3) are shown in Fig. 4.1. The computational domain size is 7 m horizontally and 12 m vertically. The distance between the transmitter and receiver is 0.2 m. The excitation source is a Ricker wavelet with a frequency of 3 kHz. The transmitter and receiver are positioned along the depth of the borehole at intervals of 0.1 m, covering a depth range of 10 m from 1 to 11 m. To capture the detailed variations of waveforms inside the borehole in the wavefield snapshots, the borehole size in the numerical simulation (0.8 m) is much larger than the actual size (0.1 m to 0.2 m). This assumption does not affect the wave components during the detection process but provides clear reflections and waveform variations within the borehole. The length of the acquired detection signal is 5 ms, with a sampling interval of 0.002 ms. The material parameters are shown in Table 3.2. Different longitudinal and transverse attenuation coefficients are used to simulate the amplitude attenuation of the longitudinal and transverse waves in the medium, where a larger attenuation coefficient indicates a more significant amplitude decay (Tables 4.1 and 4.2).

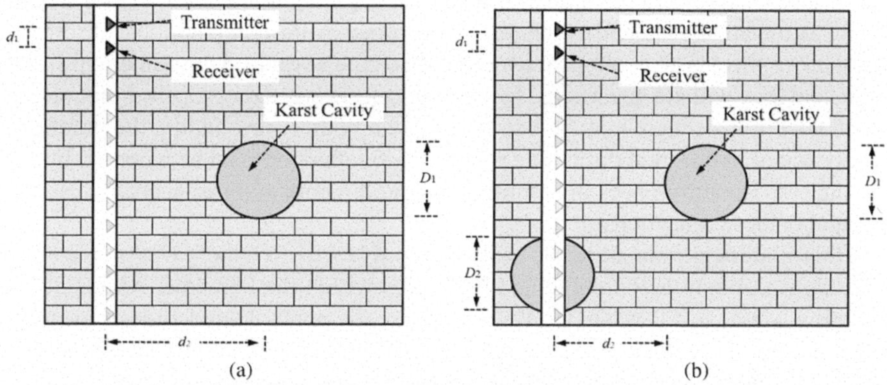


**Fig. 4.1** Scheme of borehole sonic detection simulation method **a** single void (scenario 1) **b** double void (scenario 3)

**Table 4.1** Geometry and detection parameters in simulation

| Scenarios | Spacing between transmitter and receiver $d_1$ (m) | Interval between void and borehole $d_2$ (m) | Diameter of void 1 $D_1$ (m) | Diameter of void 2 $D_2$ (m) | Wavelet frequency (kHz) |
|---|---|---|---|---|---|
| 1 | 0.2 | – | – | – | 3 |
| 2 | 0.2 | 2 | – | 2 | 3 |
| 3 | 3 | 2 | (4,0,0) | 1.6 | 3 |

**Table 4.2** Material parameters in simulation

| Material parameter | P-wave velocity $v_P$ (m/s) | S-wave velocity $v_S$ (m/s) | Density $\rho$ (kg/m3) | P-wave attenuation coefficient $\varepsilon_L$ ($s^{-1}$) | S-wave attenuation coefficient $\varepsilon_T$ ($s^{-1}$) |
|---|---|---|---|---|---|
| Bedrock | 4000 | 2300 | 2300 | 10 | 50 |
| Water in borehole | 2000 | 0 | 1500 | 1500 | 2500 |
| Infill in void | 2000 | 1100 | 1500 | 3000 | 4500 |

### 4.2.3 *Analysis for Wave Field Snapshots*

In the simulation process of Scenario 2, the velocity components are used to visualize the wavefield snapshots and analyze the detection process. Figure 4.2 shows a schematic diagram of the wavefield when detecting at a depth of 6 m in the borehole. After the horizontal excitation in the center of the borehole (as shown in Fig. 4.2 at 0.2 ms), the wavefield can be divided into three parts: waves propagating to the left through the borehole, waves propagating within the borehole, and waves propagating to the right through the borehole. The waves propagating to the left will continue to propagate toward the left and eventually be absorbed by the computational absorbing boundary, as there is no geological structure set on the left side in the numerical simulation, and they will not be received by the transducers located inside the borehole. In actual detection, as there may be karst geological bodies surrounding the borehole, waves excited in different directions will produce reflections and be recorded by the transducers. The differences in amplitude on the directional transducers are used to distinguish these reflections.

In this numerical simulation, the analysis mainly focuses on the waves propagating to the right through the borehole and the waves within the borehole. Although the excitation source is a P-wave, the P-wave within the borehole undergoes waveform conversions at the borehole. Due to the velocity difference between P-waves and S-waves in the bedrock, they gradually separate during propagation (as seen in Fig. 4.2 at 0.2 ms and 0.4 ms). Due to the faster velocity of P-waves, P-waves have already reached the near end of the cavity at 1.0 ms and appear as reflections and waveform changes in the borehole. The part where P-waves convert to S-waves is referred to as PS-waves, and the part where P-waves remain as P-waves is referred to as PP-waves. However, at moments 1.0 ms and 1.4 ms, due to significant interference between reflections and P-waves/S-waves, the reflections of PP-waves and PS-waves are not prominent. By 1.8 ms, the reflection of PP-waves has propagated to the borehole and is received by the transducers within the borehole, while the reflection of P-S waves has propagated to the middle of the borehole and the cavity, eventually reaching the borehole at around 2.0 ms. The waves transmitted into the cavity continue to propagate toward the far end of the cavity (as seen in Fig. 4.2 at 1.4 ms and 1.8 ms) and produce reflections at the far end of the cavity. The main observed reflection in

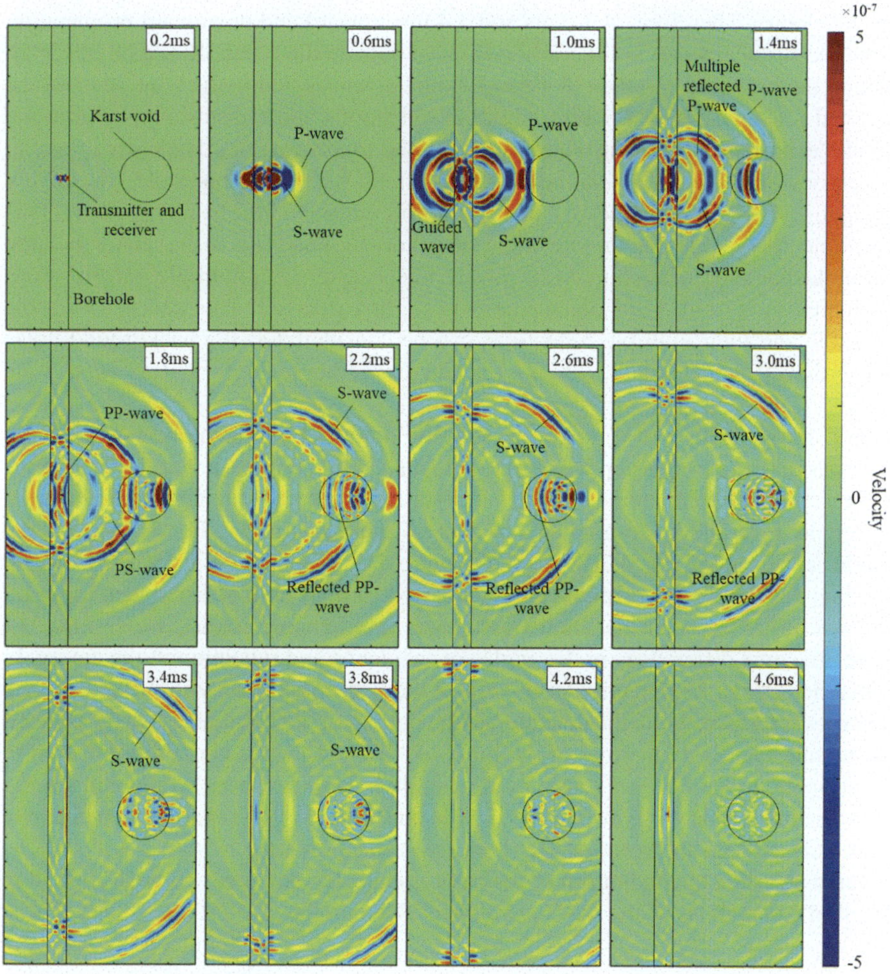


**Fig. 4.2** Snapshots of borehole sonic detection for karst void (Scenario 2, single void). The transmitter is located at the depth of 6 m

the wavefield is the PP-reflection wave at the far end of the cavity (as seen in Fig. 4.2 at 2.2 ms), which eventually propagates to the borehole at 3.2 ms.

The waves within the borehole are noteworthy in karst cavity acoustic detection, as they demonstrate two aspects: the propagation of waves within the borehole itself and the multiple reflections and transmissions of waves between the borehole and external reflections. The propagation within the borehole includes waves propagating longitudinally within the borehole, which propagate at the longitudinal wave velocity of water. After reaching the borehole wall, these waves propagate near the borehole as Rayleigh waves, which propagate at the surface wave velocity of the bedrock, which is close to the shear wave velocity and approximately 0.9 times the shear wave velocity. Therefore, at 3.0 ms and 3.4 ms, upward and downward propagating borehole surface waves similar to the S-wave outside the borehole can be observed around the borehole. On the other hand, both the waves within the borehole and the external reflected waves undergo multiple reflections and transmissions within the borehole, particularly during the excitation of the source wavelet. The source wavelet undergoes multiple reflections and transmissions within the borehole, effectively elongating the source wavelet or causing multiple emissions due to energy decay. This phenomenon is evident in the wavefield snapshot at 1.4 ms. Due to the relatively large borehole diameter in this numerical model, the occurrence of multiple transmissions compared to a small borehole scenario is slightly delayed. The waves within the borehole and their multiple reflections can be considered as the background noise in this detection because they do not contain any geological information outside the cavity.

For the simulation of a model, it actually includes emissions and receptions at intervals of 0.1 m from a depth of 1 m to a depth of 11 m, with a total of 100 emissions and receptions for each model. Each transmitting and receiving generate a different wavefield. Overall, the components of the wavefield are similar. For Scenario 2, the single cavity model, Fig. 4.2 shows the wavefield at the reflection in the center depth, and Fig. 4.3 shows the wavefield at the emission from a depth of 4 m. From Figs. 4.2 and 4.3, it can be seen that their overall wavefield characteristics are similar, but there are some differences in the angles of incidence and reflection in the cavity. In actual detection, the detection at the top of the borehole is easily influenced by the soil-rock interface, while the detection at the bottom of the borehole is easily affected by the interface at the bottom of the borehole. This influence is mainly reflected in the formation of reflection waves by the borehole surface waves propagating along the borehole wall at these interfaces. At the same time, in the scenario where a cavity is already present in the borehole, borehole reflection surface waves will also be generated at the top and bottom of the cavity. This phenomenon will be discussed in detail in the analysis of the wave profiles for Scenario 3.

### *4.2.4 Analysis for Wave Profile*

Figures 4.4, 4.5, and 4.6 respectively show the wave profiles of the horizontal component of the borehole for the three numerical simulation scenarios, composed of detection signals at different depths. The vertical axis of the wave profiles represents depth, and the horizontal axis represents time. The positive values of the signals are filled in black to better display the arrival time.

Figure 4.4 represents the wave profile of the numerical simulation for Scenario 1. In this wave profile, only the waveforms inside the borehole are included because no geological structure is set outside the borehole. The wave components inside the borehole include direct waves, borehole surface waves (pseudo-Rayleigh waves and pseudo-Stoneley waves) propagating along the borehole wall, and the reflected seismic source subwave undergoing multiple reflections inside the borehole. The

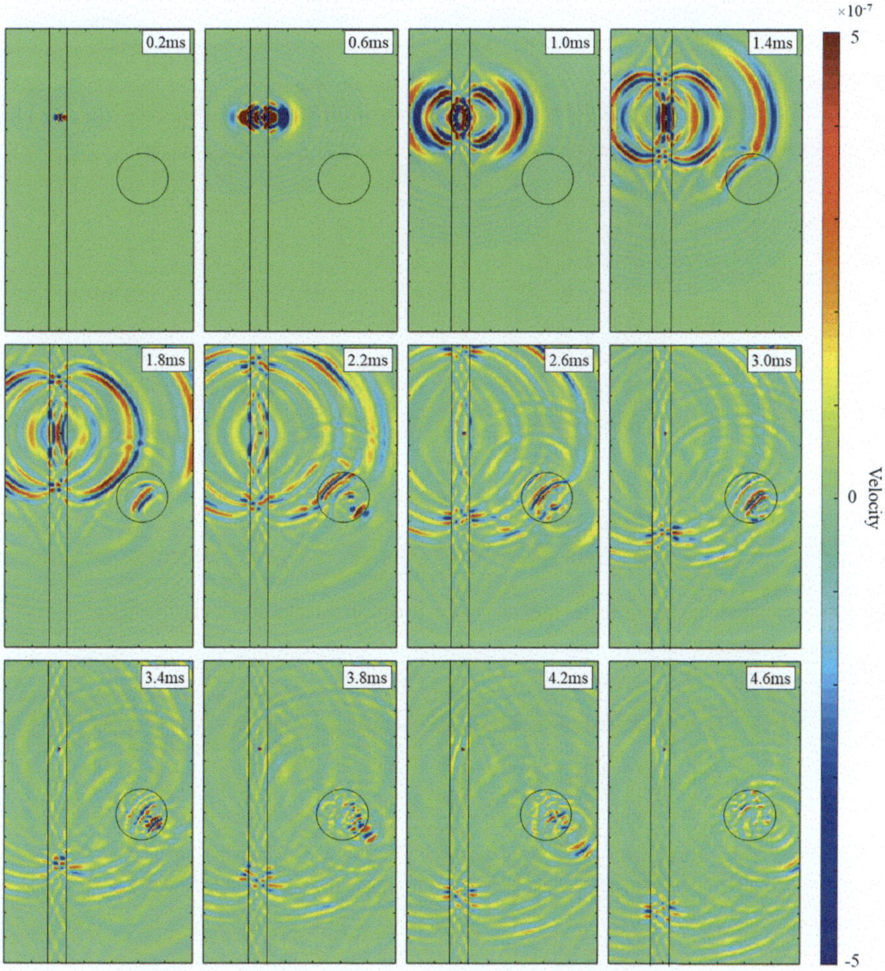


**Fig. 4.3** Snapshots of borehole sonic detection for karst void (Scenario 2, single void). The transmitter is located at the depth of 3 m

components of the waves inside the borehole are relatively complex and vary with different external geological conditions. However, it should be noted that the detection intervals are close and the distance between emissions and receptions is small, so the direct waves and the waves inside the borehole change very little over a short distance. Therefore, the wave profile mainly shows the vertical arrival time axis.

Figure 4.5 represents the wave profile of the numerical simulation for Scenario 2, the single cavity model. The wave components can be mainly divided into waves propagating inside the borehole and waves reflecting from the cavity, which is consistent with the wavefield analysis in the previous section. The components of the waves inside the borehole are similar to those in Scenario 1, the complete stratum model,

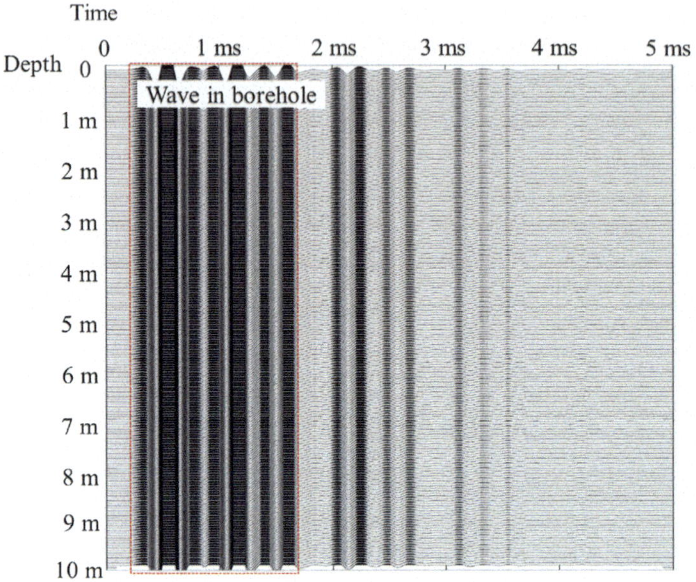


**Fig. 4.4** Simulated wave profile of borehole sonic detection (Scenario 1, intact bedrock)

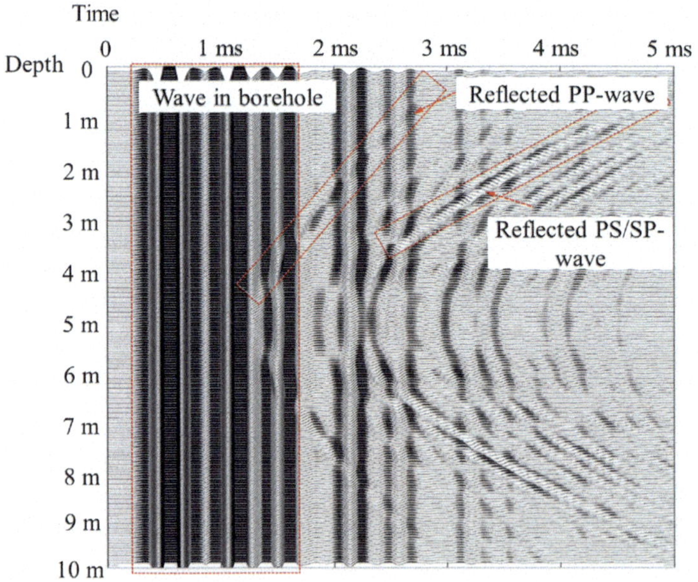


**Fig. 4.5** Simulated wave profile of borehole sonic detection (Scenario 2, single void)

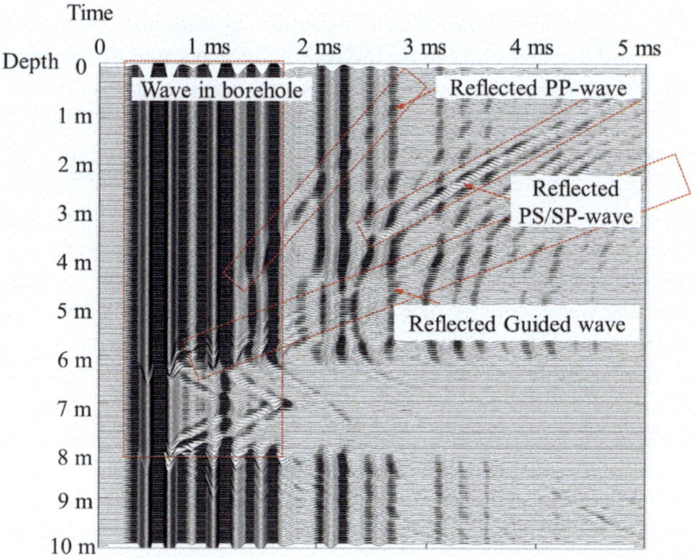


**Fig. 4.6** Simulated wave profile of borehole sonic detection (Scenario 3, the cavity encountered during drilling)

mainly consisting of direct waves and multiple reflections inside the borehole, all showing vertical arrival time axes. The reflections from the cavity are shown as hyperbolic-shaped arrival time axes in Fig. 4.5. Due to the waveform conversion of the waves at the borehole and cavity, the reflections from the near end of the cavity mainly appear as PP-waves, PS-waves, and SP-waves. PP-waves refer to waves that propagate from the borehole to the right in the form of P-waves, and after reflecting at the near end of the cavity, they continue to propagate back to the borehole as P-waves. PP-waves are the first waveforms received by the borehole transducer, and they have the steepest arrival time axis. The higher the velocity, the steeper the arrival time axis of the reflection. PP-reflection waves are reflected back to the borehole at the velocity of the P-wave in the bedrock, and their slope is close to the P-wave velocity of the bedrock. Correspondingly, there are PS- and SP-waves, which propagate from the borehole to the cavity as P-waves and S-waves, respectively, and propagate back to the borehole as S-waves and P-waves, respectively, after waveform conversion at the borehole. Since half of the distance is propagated as S-waves, their arrival time is later than that of the PP-waves, and the slope of the arrival time axis is noticeably smaller than that of the PP-waves. It can be observed that there are some reflection arrival time axes with the same slope as the PP-reflection after the arrival of the PP-reflection. This is caused by multiple reflections inside the borehole.

Figure 4.6 represents the wave profile of the numerical simulation for Scenario 3, where the borehole has encountered a cavity. In actual engineering, encountering a cavity in the borehole is a common scenario. When the borehole has encountered a cavity, the detection of the cavity outside the borehole becomes more complex. The purpose of this numerical simulation in this scenario is to identify the influence of the cavity encountered by the borehole on the reflection wavefield of the cavity outside the borehole. Compared to the previous two scenarios, the differences in the wave profile shown in Fig. 4.6 can be mainly divided into three parts: (1) The generation of reflection surface waves from the borehole already encountering the cavity. These surface waves are generated at the interface between the borehole wall and the top and bottom of the cavity and propagate along the borehole wall. The arrival time axis of the borehole surface waves is shown in Fig. 4.6, and its slope is smaller than that of the PS- and SP-waves. Since the propagation form of borehole surface waves is surface waves, the wave velocity of surface waves is significantly lower than that of longitudinal waves and slightly lower than that of transverse waves. (2) In the depth range where the borehole has encountered the cavity, the multiple reflections of the borehole are not significant because the medium filling the cavity has a significant attenuating effect on the reflected waves. This attenuation also affects the reflected waves from the cavity outside the borehole received by the transducer at that depth. It can be observed that within this depth range, the reflection arrival time axes from the cavity outside the borehole appear discontinuous. (3) Within the depth range where the borehole has encountered the cavity, there are reverse hyperbolic reflection arrival time axes due to the boundary of the cavity. In actual scenarios, since cavities are mostly not point-like circles in shape, these types of reflection arrival time axes may not necessarily be hyperbolic.

## 4.3 Signal Processing Method

The detection of cavities around the borehole utilizes two types of acoustic transducers with different frequencies and four directional receivers to assess the development of cavities around the borehole. The basic idea is to use high-frequency signals (tens of kHz) to identify rock fractures and cavities encountered by the borehole, and to use low-frequency signals (thousands of Hz) to assess the development of cavities in the vicinity of the borehole. Since the emission and reception distances are very close, the detection profiles can be considered as near-zero-offset profiles. The signal processing flow for near-zero-offset profiles is shown in Fig. 4.7 and can be divided into two main parts: (1) The processing of high-frequency signals involves frequency filtering and signal energy calculation. (2) The processing of low-frequency signals includes frequency filtering, median filtering, and frequency-wavenumber (F-K domain) filtering to extract the reflection waves from the cavities outside the borehole. The analysis results of high-frequency signal processing contribute to the parameter selection for F-K domain filtering of the low-frequency signals. In this section, we will provide a detailed analysis of signal energy calculation, median filtering, and F-K domain filtering.

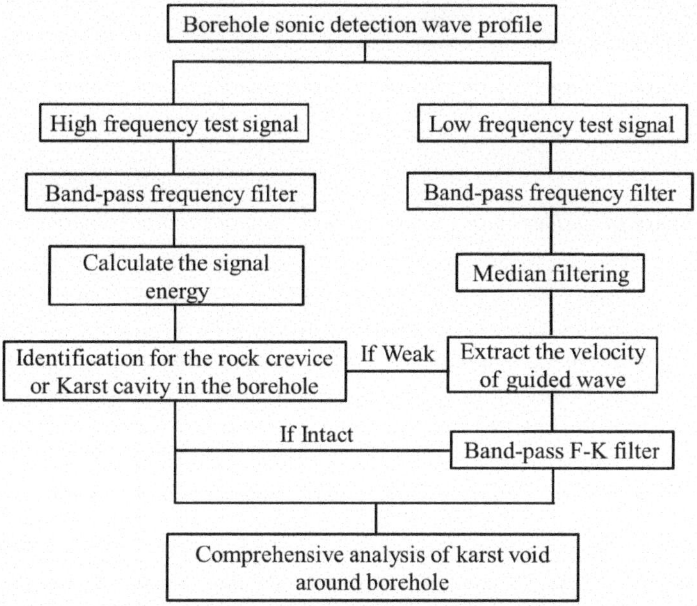


**Fig. 4.7** Signal processing workflow of borehole sonic detection of karst void

### *4.3.1 Signal Energy Calculation*

The signal energy of every trace characterizes the integrity of the rock at the borehole wall. The signal energy can be calculated as follows:

$$E_z = \int_{t_0}^{t_1} s_{hz}^2(t)dt \tag{4.28}$$

where $E_z$ is the signal energy at a depth of $z$. $S_{hz}$ $(t)$ is the high-frequency test signal at a depth of $z$. $t_1$ and $t_0$ are the start and end of the test signal.

The energy strength of a certain depth detection signal $S_{hx}$ $(t)$ is represented by the grayscale color bar shown in Fig. 4.8, which reflects the integrity of the borehole wall rock. When the color tends to white, the borehole wall rock is intact. When the color tends to black, the encountered formation may be fractured rock or a cavity. The colors of signals at different depths will form an illustration of the changing integrity of the borehole wall with depth.

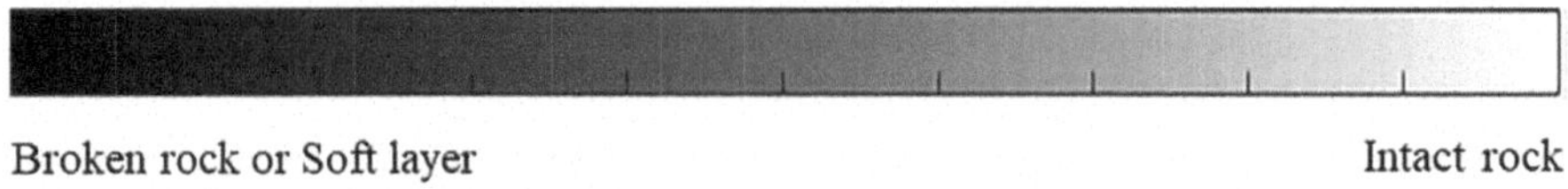


**Fig. 4.8** Evaluation of rock intactness of borehole wall using a gray color bar

### *4.3.2 Medium Filter*

Based on the analysis of the wavefield and detection waveforms, it is known that the original low-frequency detection signal waveform $P_l(t, x)$ mainly includes direct waves, multiple reflections within the borehole, and reflections from geological formations outside the borehole. Due to the small distance between the transmitting and receiving transducers and the short distance between adjacent source points, we can assume that the direct waves and multiple reflections within the borehole are consistent within a certain adjacent depth range. In this small depth range, the variations of the direct waves and multiple reflections within the borehole can be neglected. Based on this assumption, we consider the same-phase axes of the direct waves and multiple reflections within the borehole to be vertical, while the reflection same-phase axes of geological formations outside the borehole generally have certain slopes, as the travel time of the reflection waves from geological formations outside the borehole varies with the positions of the transmitting and receiving transducers. By using the differences in the same-phase axes between the effective reflection waves (reflections from geological formations outside the borehole) and the background noise (direct waves and multiple reflections within the borehole), we can use median filtering to differentiate between these two types of waveforms. The specific method is as follows:

(a) Take the test signal at one depth $S_x(t)$ as the center; a new set of sequences can be formed by reading the test signals of $2N + 1$ channels at adjacent depths $(S_{x\text{-}N}(t), S_{x\text{-}N+1}(t)\ldots S_{x+N}(t))$.
(b) For every point at time $S_x(t_n)$, the amplitudes of the test signals of $2N + 1$ channels $(S_{x\text{-}N}(t_n), S_{x\text{-}N+1}(t_n)\ldots S_{x+N}(t_n))$ are reordered. The median $S_{xm}(t_n)$ replaces the original amplitude $S_x(t_n)$ in the test signal $S_x(t)$.
(c) Repeat steps (a) and (b) using the test signal at the next depth $S_{x+1}(t)$. A new wave profile after median filtering can be obtained as $P_m(t, x)$, which consists of the direct wave and multiple reflected waves in the borehole.
(d) The wave profiles $P_r(t, x)$ that contain the reflected waves from the geological structure can be calculated as $P_\text{r}(t, x) = P_l(t, x)\text{-} P_\text{m}(t, x)$.

Based on the median filtering method described above, the waveforms of the single cavity condition (Fig. 4.5) and the borehole-cavity condition (Fig. 4.6) were subjected to median filtering, and the results are shown in Fig. 4.9. Figure 4.9a and b represent the median filtering results for the single cavity condition, where Fig. 4.9a shows the filtered waveform and Fig. 4.9b shows the removed waveform. The sum of Fig. 4.9a and b corresponds to the original waveform shown in Fig. 4.5. After median

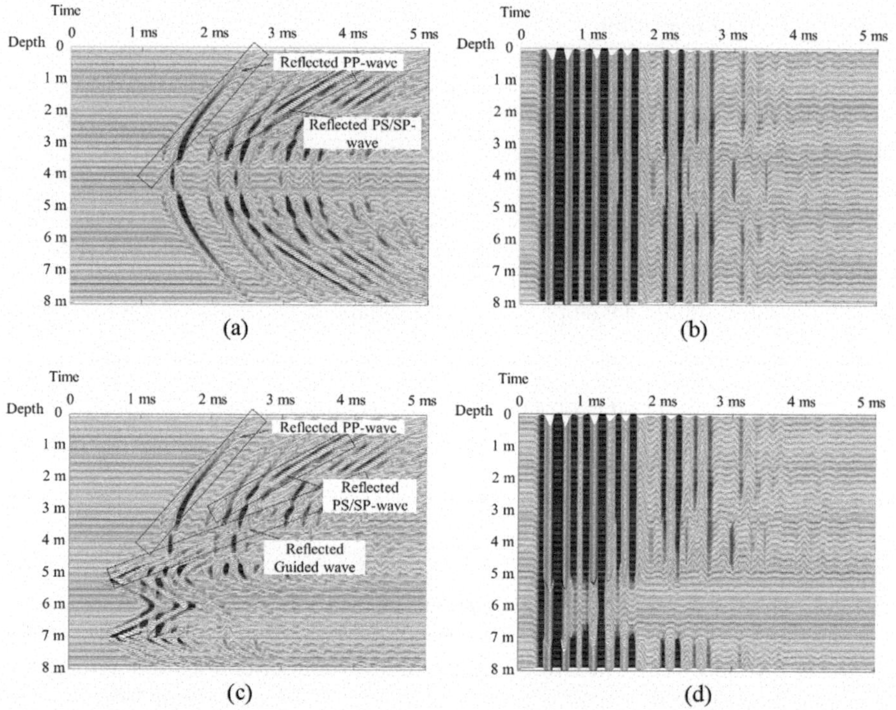


**Fig. 4.9** Median filter of simulated wave profile of borehole sonic detection **a** median filtered profile (scenario 2). **b** Removed wave profile using median filtering (scenario 2). **c** Median filtered profile (scenario 3). **d** Removed wave profile using median filtering (scenario 3)

filtering, the effective reflection signals (Fig. 4.9a) and the background noise signals within the borehole (Fig. 4.9b) are completely separated. Figure 4.9c and d show the median filtering results for the borehole-cavity condition, where the effective signals (Fig. 4.9c) and the background noise (Fig. 4.9d) are also separated through median filtering. The separated background noise (Fig. 4.9b and d) does not exhibit same-phase axes with slopes. In the median filtering process, N = 10, so the top and bottom 10 traces of the waveform are removed, and the actual analysis is conducted for depths ranging from 2 to 10 m.

### 4.3.3 F-K Filter

The F-K filter (Duncan and Beresford 1995) of the wave profile $P_{lr}$ $(t, z)$ is used to identify the reflected P-wave based on the wave velocity difference between the body wave, converted wave, and Stoneley wave. $P_{lr}$ $(t, z)$ in a time–space domain can be converted to the frequency-wavenumber domain $G(\omega, k_z)$ by a two-dimensional frequency Fourier transform:

$$G(\omega, k_x) = \int_{-\infty}^{\infty} \int_{-\infty}^{\infty} P_{lr}(t, z) e^{-j(\omega t + k_x x)} dtdz \tag{4.29}$$

where $k_x$ is the wavenumber (i.e., the reciprocal of the wavelength). The circular frequency $\omega = 2\pi f$. Thus, the wave velocity $v = f/k$ is the slope in the F-K domain, as shown in Fig. 4.10. The reflected P-wave is of relatively higher velocity so that its event has a more significant slope in the F-K domain. A bandpass velocity filter can be designed as follows:

$$G_s(\omega, k) = G(\omega, k) \cdot H(\omega, k) \tag{4.30}$$

$$H(\omega, k) = \begin{cases} 1, & v_2 \leq \left|\frac{f}{k}\right| \leq v_1 \\ 0, & \left|\frac{f}{k}\right| \leq v_2 \text{ or } \left|\frac{f}{k}\right| \geq v_1 \end{cases} \tag{4.31}$$

where $H(\omega, k)$ is a velocity filtering mask. The wave velocity between $v_1$ and $v_2$ remained after filtering, as shown in the shaded area in Fig. 4.10. $G_s(\omega, k)$ is the processed F-K spectrum. The two-dimensional inverse Fourier transform returns the processed data to the time–space domain as $R(t, z)$. Then, the reflected P-wave can be observed.

$$R(t, z) = \frac{1}{2\pi} \int_{-\infty}^{\infty} \int_{-\infty}^{\infty} G_s(\omega, k) e^{j(\omega t + kz)} dtdk \tag{4.32}$$

The waveforms shown in Fig. 4.9a and c were subjected to F-K domain filtering, using a passband filtering window centered around the longitudinal wave velocity. The filtering results are shown in Fig. 4.11a and b, where PS- or SP-reflection waves and borehole surface waves are filtered out, while the portion that propagates in the wavefield in the form of P-waves is retained. The multiple reflections within the borehole caused by PP-reflection waves can still be observed, but these waveforms do not affect the identification of the travel time of the PP-reflection waves. Overall, F-K domain filtering simplifies the waveform components, allowing for the extraction of specific slope same-phase axes as needed.

## 4.4 Validation

In an experimental setup conducted indoors, the effectiveness of high-frequency acoustic wave detection was assessed within a borehole. The evaluation encompassed the following controlled tests: (1) determining the ability to detect broken rock or

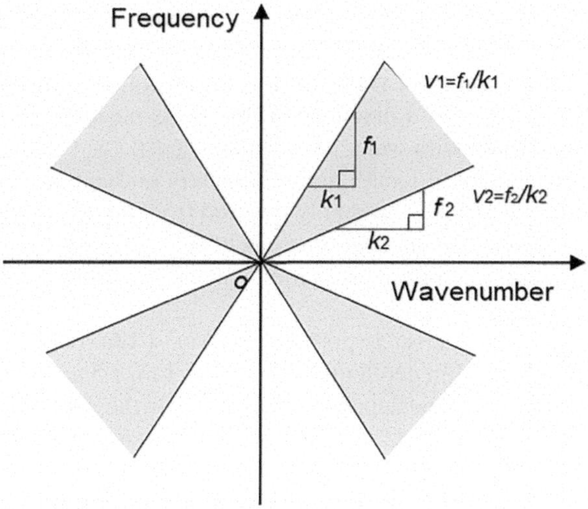


**Fig. 4.10** Scheme of F-K filer for borehole low-frequency sonic data

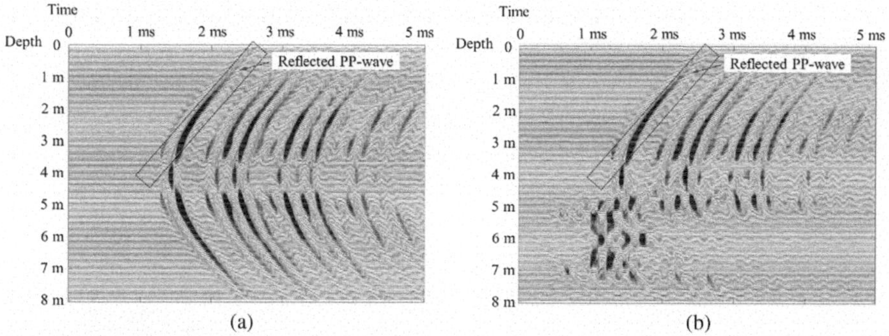


**Fig. 4.11** F-k filter of simulated wave profile of borehole sonic detection **a** F-K filtered profile of scenario 2. **b** F-K filtered profile of scenario 3

karst cavities filled with water or sandy soil using high-frequency acoustic waves, (2) assessing the performance of azimuthal receivers in detecting low-frequency acoustic waves, and (3) analyzing the variations in the test signals obtained by the azimuthal receivers.

### *4.4.1 High-Frequency Detection Tests*

Figure 4.12 depicts the experimental arrangement employed to test the high-frequency acoustic transmitter. It consists of a concrete pile containing three precast cylindrical voids filled with broken rock, sandy soil, and water. The broken rock has a rock quality designation (RQD) of 20, while sandy soil and water represent typical fill materials in a karst cavity. The voids measure 0.5 m in length and have a diameter of 0.4 m. A PVC casing with a diameter of 90 mm is positioned in the center of the concrete pile and filled with water to facilitate acoustic wave coupling. The pile itself measures four meters in length and two meters in diameter. Throughout the experiment, an acoustic probe is gradually lowered from the top of the PVC tube to the bottom at a velocity of 0.1 m/s. At intervals of 0.1 m, a high-frequency acoustic wave is transmitted using a constant excitation voltage and subsequently detected by four azimuthal receivers.

The wave profile of a receiver is presented in Fig. 4.13a, revealing distinct and continuous wave events within depth intervals of 0 to 1 m, 1.5 to 2.0 m, 2.5 to 3.0 m, and 3.5 to 4.0 m. However, in the depths corresponding to the voids filled with broken rock, sandy soil, and water, the wave events become challenging to identify, resulting in weak signals. To represent the rock integrity near the borehole, a grayscale color bar is utilized to illustrate the energy of the test signal. When the color tends toward white, it indicates intact rock, whereas colors closer to black suggest broken rock or the presence of a soft layer within the stratum. The energy of the signal is highest in the intact depth range, followed by the broken rock range (1 to 1.5 m), sandy soil range (2 to 2.5 m), and water depth range (3 to 3.5 m). The energy of the signal in

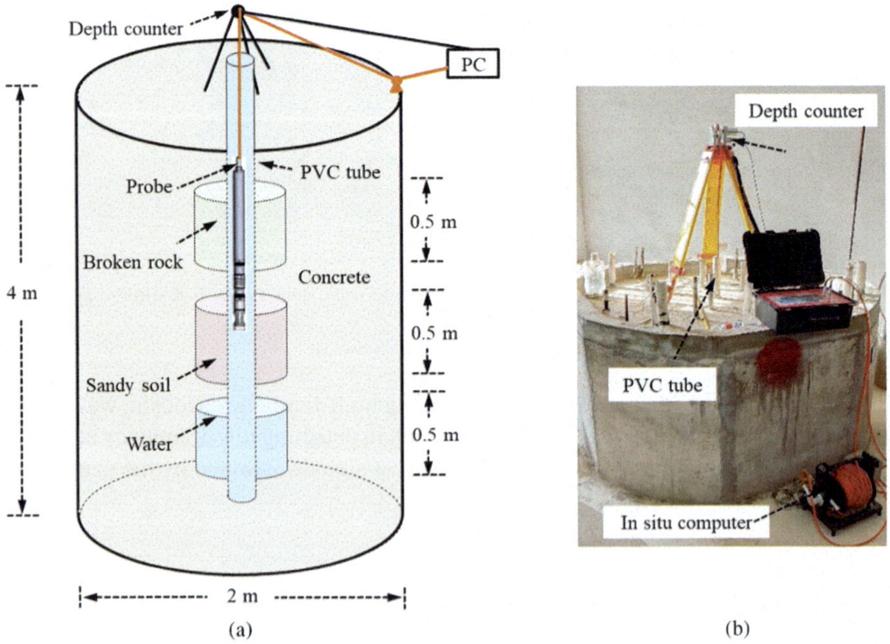


**Fig. 4.12** Borehole experimental setup for the high-frequency detection test

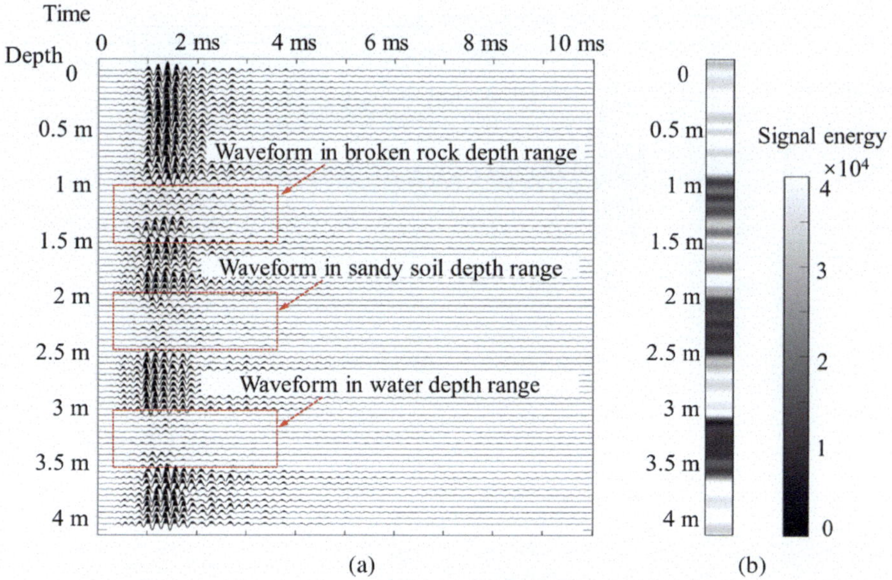


**Fig. 4.13** High-frequency acoustic wave detection results: **a** test wave profiles, **b** energy imaging results

the 1.3 to 1.4 m range is similar to that of the intact range, owing to the presence of some intact rock within the broken rock range.

Figure 4.14 presents the test signals and their spectra at depths of 0.5 m, 1.25 m, and 2.25 m. A notable difference can be observed in the amplitude of the test signal between the intact concrete and the broken rock, sandy soil, and water. The corresponding spectra reveal that the test signal in the intact concrete exhibits higher energy and broader bandwidth compared to the other signals (indicated by points A, B, C, and D in Fig. 4.14b). This finding confirms the effectiveness of the high-frequency acoustic test signal's energy as an indicator for identifying broken rock or a karst cavity along the borehole.

### 4.4.2 *Azimuthal Receiver Tests*

To evaluate the performance of the azimuthal receivers, a water tank experiment was conducted following the setup depicted in Fig. 4.15. The experiment described by Che et al. in their works from 2014 and 2017 was considered suitable for this purpose. In the experiment, a cylindrical water tank with a diameter of 3 m was used, and a PVC tube was positioned at the center. The water inside the tank had a density of approximately 1000 kg/m$^3$ and an acoustic velocity of 1500 m/s. The water depth was set at 4 m. Within the tank, a metal block representing an imaging target was placed at a depth of 2 m and a distance of 1 m from the PVC tube. The dimensions of

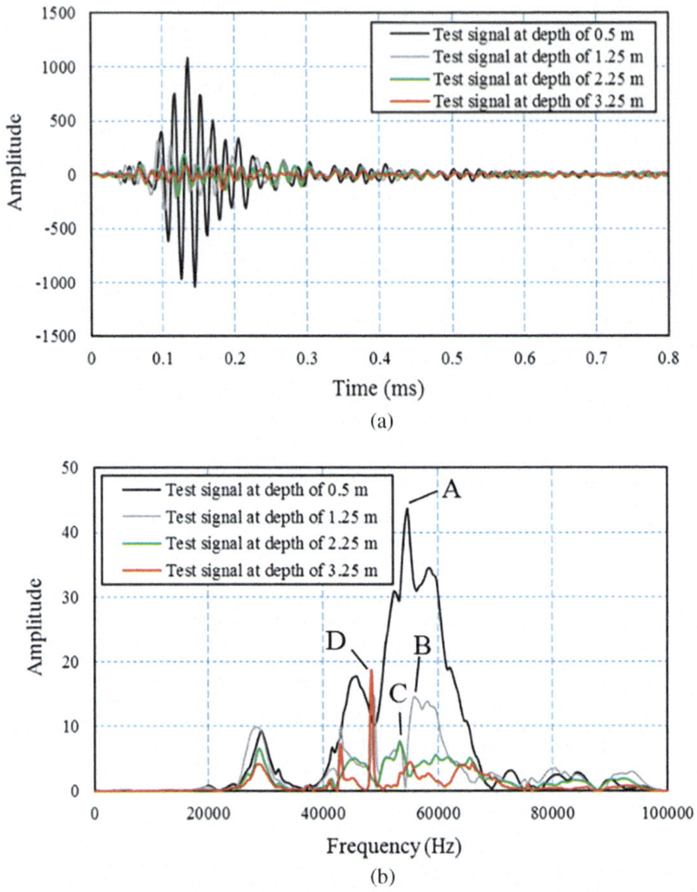


**Fig. 4.14** **a** Test signals at different depths and **b** their corresponding spectra

the metal block were 0.3 m × 0.3 m × 0.3 m. Low-frequency acoustic wave detection was conducted by sequentially moving from the top to the bottom of the PVC tube at 0.1 m intervals. The test signal was simultaneously received by all four azimuthal receivers, with Receiver No. 1 facing the metal block, as shown in Fig. 4.15.

Figure 4.16a presents the wave profile obtained from the water tank experiment without a metal block. The profile displays various waves, including those within the PVC tube, multiple reflections from the tank wall, and interface-reflected waves from the top and bottom of the tube. Certain waves, such as those in the borehole and multiple reflections, exhibit constant arrival times (indicated by the vertical red dotted box in Fig. 4.16a), allowing for their removal using median filtering. Figure 4.16b showcases the wave profile with a metal block positioned at a depth of 2 m and 1 m away from the PVC tube. In comparison to the other arrivals in Fig. 4.16a, the reflected wave from the metal block is relatively weak. Nonetheless, by employing

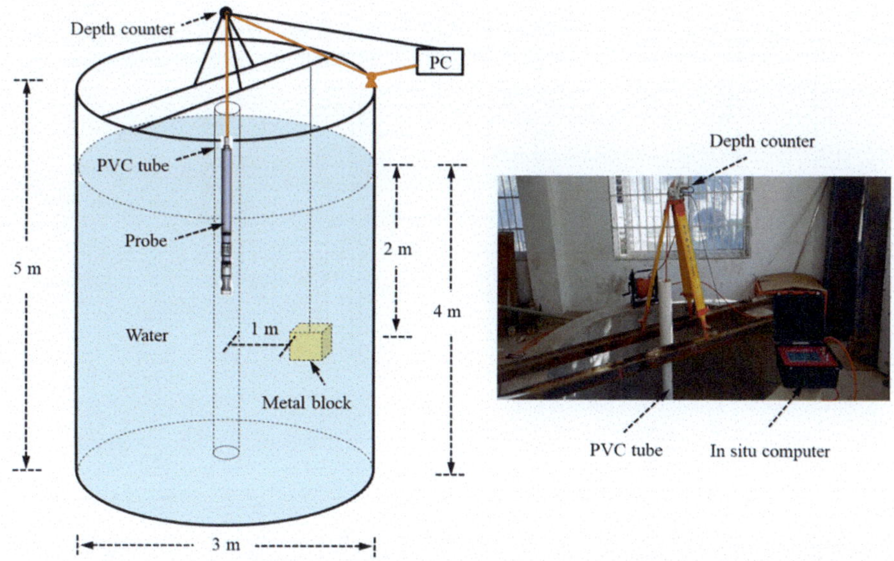


**Fig. 4.15** Experimental setup for the azimuthal acoustic receiver performance test

median filtering (as demonstrated in Fig. 4.16c and d, it is possible to identify the presence of the reflected metal block. In this particular test, a parameter N of three is used for the median filtering. The constant arrival time events are effectively filtered out, as displayed in Fig. 4.16c. Meanwhile, the hyperbolic-shaped reflected wave from the metal block and the guided reflected wave remains visible in Fig. 4.16d, enabling the identification of objects near the borehole.

Figures 4.17 and 4.18 display the original test signals, median filtered signals at a depth of 2 m, and their respective spectra obtained from rotating the downhole probe at twelve azimuths. In Fig. 4.17a, small differences can be observed at 1 ms and 4 ms (indicated by the dotted black box), although these differences are not significant. Consequently, the reflected wave amplitude differences among the azimuthal receivers are difficult to identify in the original test wave signals. However, after applying median filtering, the amplitude differences in the signals obtained by the azimuthal receivers become more pronounced, as shown in Fig. 4.18a. This difference in amplitudes is also visible in the corresponding spectra, as depicted in Figs. 4.17b and 4.18b. Overall, median filtering enhances the differentiation of main frequency amplitudes among the azimuthal receivers.

Figure 4.19 presents a comparison between the overall energy of the test signal and the energy of the reflected wave based on the receiver azimuth and the presence of a metal block. The reflected wave is derived by subtracting the test signal without the metal block from the signal with the metal block. Notably, there is minimal disparity in the total energy of the test signals across various azimuths (as indicated by the black line in Fig. 4.19). However, a substantial difference in the energy of the reflected wave can be observed among different azimuths (as illustrated by the red line in Fig. 4.19).

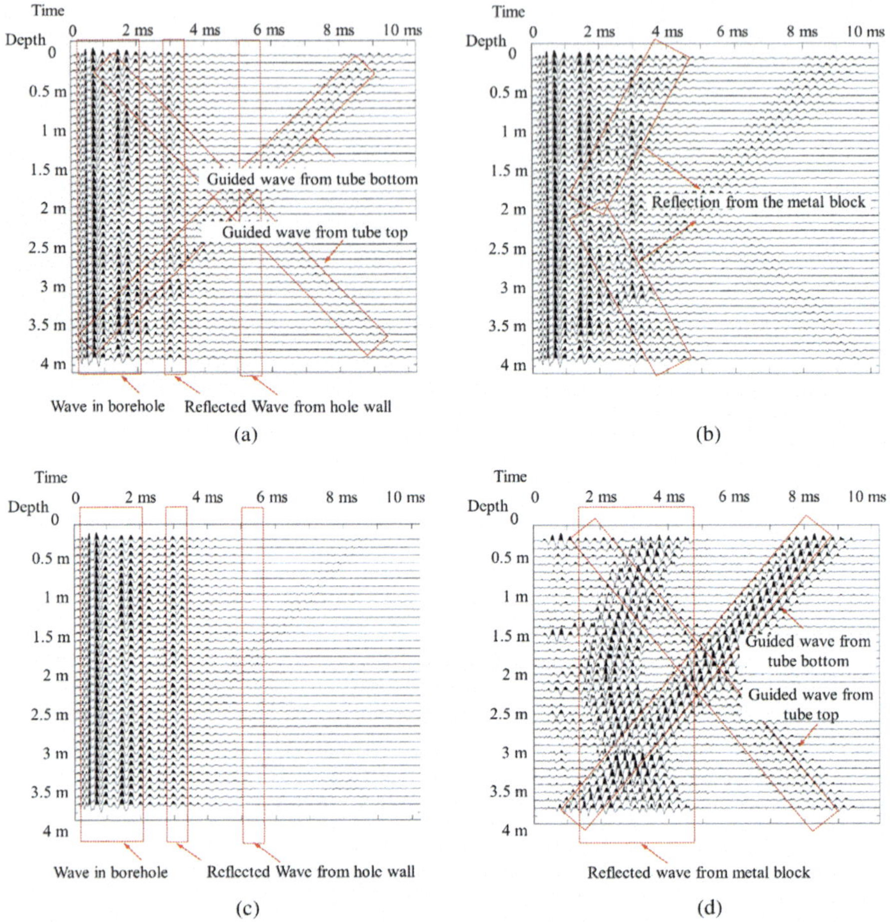


**Fig. 4.16** Low-frequency acoustic wave detection results of the no.1 receiver. **a** Test wave profiles without a metal block in the water tank. **b** Test wave profiles with a metal block at a depth of 2 m. **c** Computed median profile with a metal block. **d** Median filtering results with a metal block after subtracting profile **c** from profile **b**

This finding confirms the effectiveness of azimuthal receivers in determining the direction of the reflected signal. Specifically, the receiver facing the metal block exhibits a greater amplitude in the reflected wave compared to the receivers in other azimuths. Consequently, utilizing azimuthal receivers in conjunction with median filtering proves viable for identifying karst cavities near the borehole.

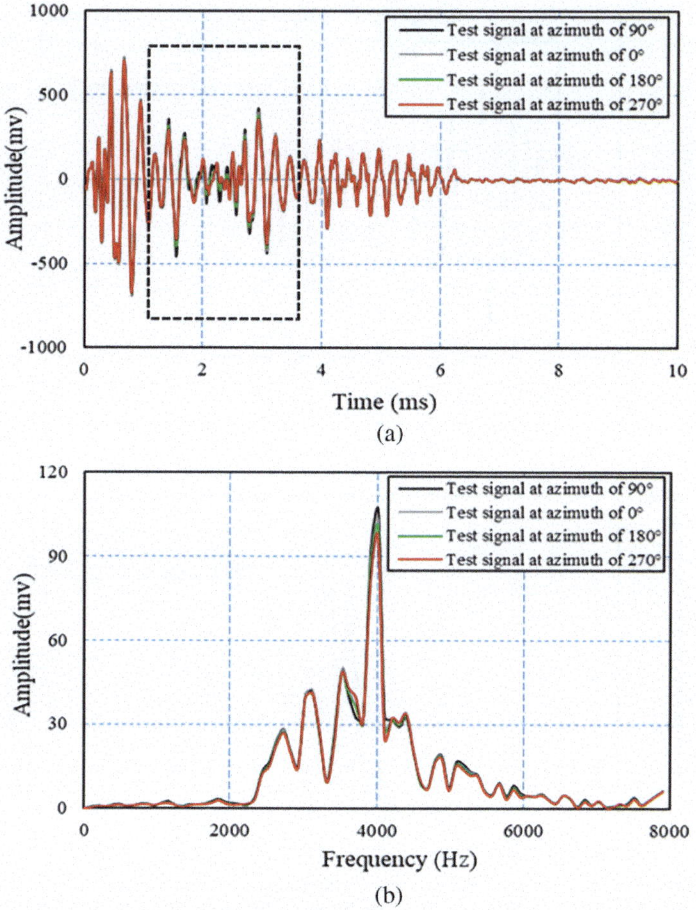


**Fig. 4.17** **a** Original test signals at a depth of 2 m at different azimuths and **b** their corresponding spectra

# 4.5 Case Study

## *4.5.1 Construction Background*

The Qingxi Bridge, depicted in Fig. 4.20, is situated in Qingyuan city, Guangdong Province, China. Spanning across the Bei River, it serves as a vital connection between the Guangqing Expressway and Qingxi Expressway. Table 4.3 provides an overview of the simplified stratigraphy observed at the site. The overlying layers consist of Quaternary deposits with an approximate thickness of 25 m. These deposits encompass plain fill, silty clay, fine sand, pebbles, and medium sand. The underlying bedrock belongs to the Devonian Donggang Formation ($D_2d$) and primarily

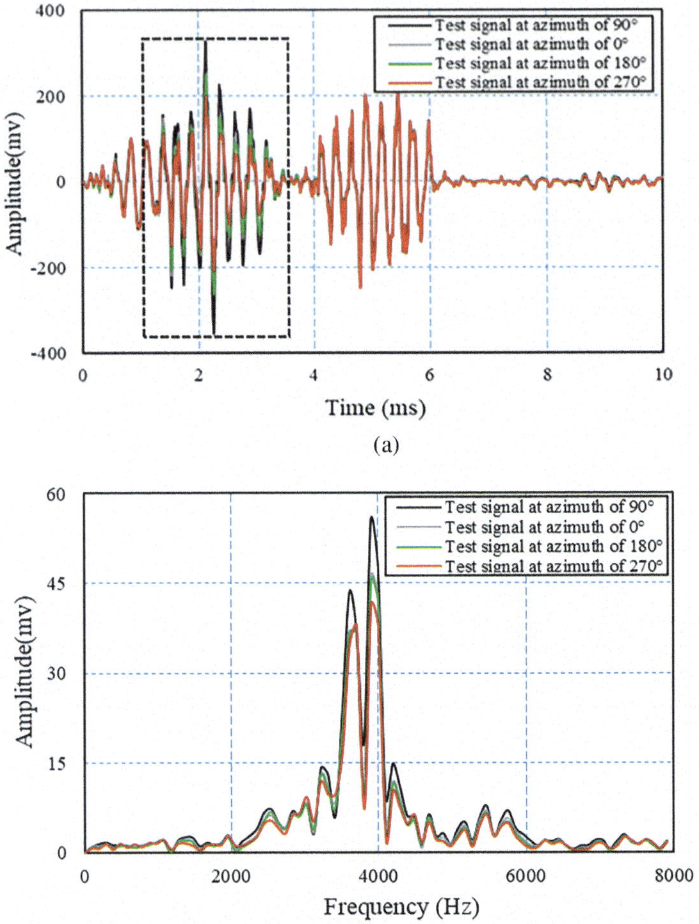


(b)

**Fig. 4.18** **a** Median filtered signals at a depth of 2 m at different azimuths and **b** their corresponding spectra

comprises medium-weathered limestone in the upper layer, characterized by a rock quality designation (RQD) exceeding 40. The lower layer consists of weakly weathered limestone with an RQD exceeding 60. Groundwater occupies the joints and fissures within the bedrock, with limited water retention due to the presence of jointing and mud-filled crevices. The limestone is also filled with carbonate karst fissure water, which demonstrates good permeability owing to the predominant presence of sand, gravel, and pebbles as fill materials within the karst formations. Notably, karst cavities are encountered in almost every borehole within the pile location, with the largest cavity boasting a diameter of 21.4 m. Braided karst cavities are commonly observed in the project area.

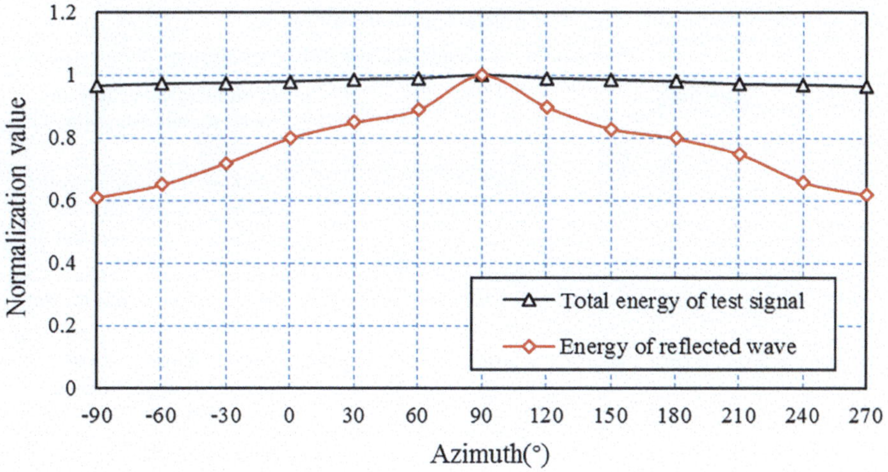


**Fig. 4.19** Maximum value of imaging using the data obtained in the different azimuths

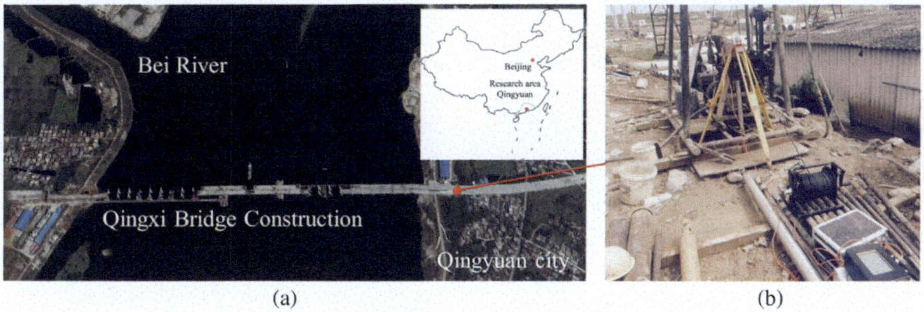


**Fig. 4.20** Location of the studied area (Qingxi Approach Bridge) in southern China in the city of Qingyuan. **a** Simplified map. **b** Synthetic photo of the field experiment

## *4.5.2 Field Test*

A field test was carried out on the Qingxi Approach Bridge to evaluate the performance of the multifrequency wave system, as depicted in Fig. 4.20. For each pile location in the project, at least one borehole was drilled. Figure 4.21 illustrates the drilling cores obtained from borehole ZK-80. The overburden in the test area comprises fine and medium sand with a thickness of 20 m. From a depth of 20 to 23 m, medium-weathered limestone is observed. Between depths of 23 to 25 m, a water-filled karst cavity is present. Weakly weathered limestone is identified in the depth range of 25 to 35 m. Based on the intact rock observed below 30 m in the drilling core, the pile bearing stratum was designed at approximately 30 m in depth. The MFAW detection technique was employed on borehole ZK-80 to detect any

**Table 4.3** Simplified stratigraphy of the Qingxi approach bridge project

| | | Formation | Average thickness (m) | Description |
|---|---|---|---|---|
| Cenozoic | Quaternary | – | 2.5 | Plain fill: gray yellow, dry, loose, mainly consisting of fine sand |
| | | – | 3.5 | Medium sand: brown yellow, saturated, loose, suitable gradation, small amount of clay |
| | | – | 6.0 | Silty clay: gray yellow, wet, plastic, cohesive |
| | | – | 3.7 | Fine sand: gray yellow, saturated, slightly dense, poor gradation |
| | | – | 6.3 | Pebbles: gray yellow, saturated, dense, suitable gradation, made of sandstone |
| | | – | 3.5 | Medium sand: gray yellow, saturated, dense, suitable gradation |
| Phanerozoic | Devonian | Donggang | 25 | Medium-weathered limestone: gray-white, cryptocrystalline structure, layered structure, developed joints and fractures, RQD is larger than 40 |
| | | | Larger than 15 | Weakly weathered limestone: gray, cryptocrystalline structure, layered structure, developed joints and fractures, RQD is larger than 60 |

karst cavities at the pile location and determine the appropriate pile bearing stratum and construction method.

The test results for receiver no. 1 with an azimuth of 37 degrees are presented in Fig. 4.19. Figure 4.22a displays the high-frequency detection wave profile along with the corresponding color bar representing signal energy at different depths. Weak wave signals are observed in the ranges of 0 to 20.1 m, 23.3 to 25.2 m, 32.2 to 32.8 m, and 33.1 to 33.4 m (indicated by the red dotted box). The interface between the overburden and limestone is identifiable at a depth of 20.1 m, consistent with the drilling core findings. The karst cavity identified in the drilling core corresponds to the region of weak energy between 23.3 to 25.2 m. However, the color bar indicates that the rock near 33 m is broken, which is not reflected in the drilling core. This discrepancy arises from the limitations of drilling core precision due to the integrity of rock and sediment within the borehole. Broken rock and soil are challenging to obtain during

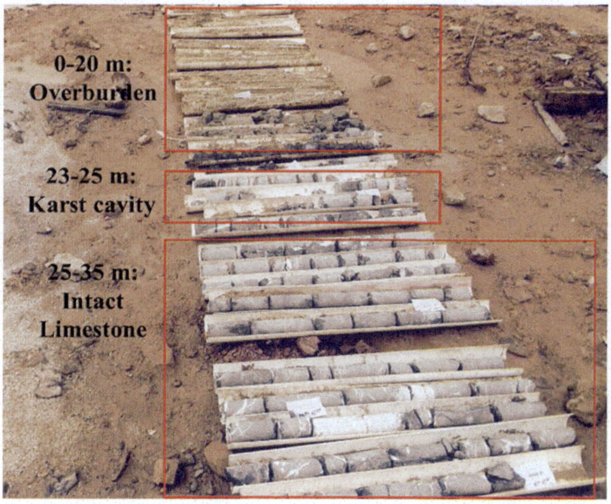


**Fig. 4.21** Drilling cores of borehole ZK-80

drilling, and the presence of unknown thick sediments in the overburden can affect the accuracy of drilling core depth.

Figure 4.22b illustrates the corresponding low-frequency detection wave profile. Discontinuities are observed in the direct wave events as the wave energy changes with depth (notably from 32.2 to 32.8 m and 33.1 to 33.4 m, indicated by the red dotted boxes). Two weak sloping events are discernible, identified as guided reflected waves, as their slopes can estimate a wave velocity of approximately 1000 m/s (e.g., using point A and point B in Fig. 4.22b). These guided reflected waves are generated at the interfaces with different impedance along the PVC tube and the borehole wall. Combining the analysis results from high-frequency detection, the reflected interfaces correspond to the karst cavity roof at 23 m and the rock fracture at 33 m. However, the significance of the guided reflected waves suggests that these waves may be disregarded in practical applications. The irregularity of reflected wave events arriving at later times than the direct waves in those depth ranges suggests the presence of several reflective geological features.

To mitigate the interference caused by direct waves, median filtering is employed to identify geological features away from the borehole. The original wave profile (Fig. 4.22b) is divided into two profiles: a calculated median wave profile (Fig. 4.23a) mainly consisting of direct waves and waves within the borehole, and a median filtered wave profile (Fig. 4.23b) primarily comprising reflected waves. The sloping reflected guided waves become more prominent after filtering. The reflected wave exhibits a chaotic appearance in the depth range where the borehole intersects the karst cavity (highlighted by the red circle in Fig. 4.23b) due to the complex fill within the karst cavity. Notably, a hyperbolic reflected wave event is observed in the depth range of 26 to 31 m. The slope of the hyperbolic event is steeper than that of the guided waves, indicating a higher velocity of the reflected wave compared to the guided wave. This suggests that the reflected wave propagated as a body wave, and the hyperbolic reflected wave event is attributed to a karst cavity near the borehole.

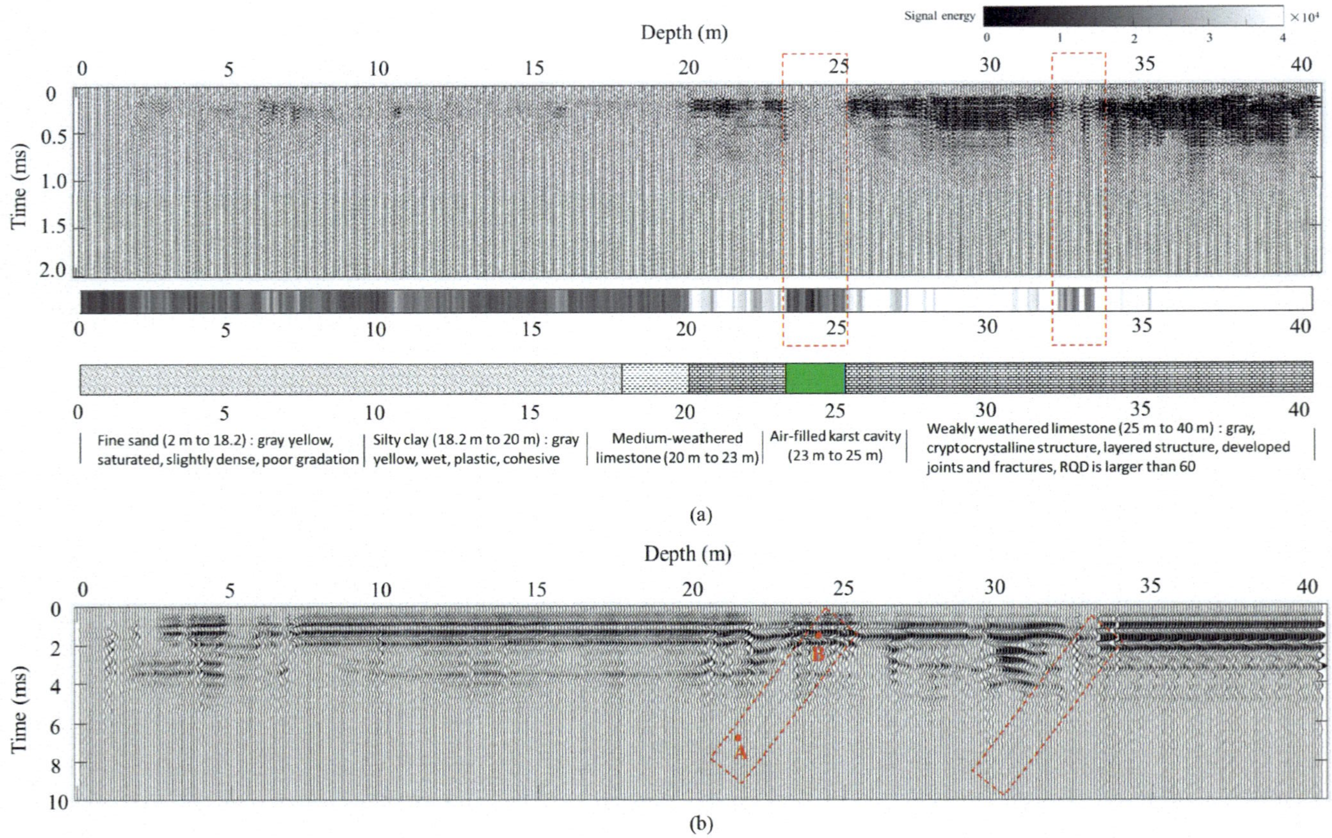


**Fig. 4.22** Test results of the no. 1 receiver whose azimuth is 37 degrees: **a** high-frequency detection wave profile and corresponding imaging color bar, **b** low-frequency detection wave profile

Using the arrival time (2.8 ms) of the reflected wave and an estimated wave velocity (2500 m/s), the distance between the karst cavity and the borehole is estimated to be approximately 3 to 4 m. The azimuthal profiles of the four receivers are utilized to determine the azimuth of the karst cavity. The median filtered azimuthal test signal at a depth of 27.8 m is extracted and analyzed (Fig. 4.24). The amplitude of the full waveform test signal is largest at an azimuth of 34 degrees and relatively smaller at azimuths of 214 and 304 degrees. When examining the first peak amplitude at approximately 2.3 ms, the signals at azimuths of 34 degrees and 124 degrees are nearly identical. However, the signal at an azimuth of 34 degrees remains relatively stronger compared to the others afterward. This variability and uncertainty may be attributed to the irregular boundary of the karst cavity near the borehole. Based on this study, the azimuth of the karst cavity near the borehole is tentatively identified as Northeast.

To validate the aforementioned detection analysis result, a borehole is drilled northeast of borehole ZK-80 at a distance of 3 m. The drilling core from 25 to 30 m confirms the presence of a karst cavity within the depth range of 27.4 to 28.7 m (indicated by two red boxes in Fig. 4.25). The karst cavity is filled with small broken limestone fragments.

In summary, the MFAW detection results indicate the presence of a karst cavity near the pile location at a depth of 28 m and a significant rock fracture at a depth of 33 m. Therefore, it is recommended to deepen the advance drilling boreholes and conduct further testing. The detection wave profile from 35 to 40 m reveals higher rock integrity at a depth of 35 m and no significant reflector outside the drilling borehole. Hence, the pile bearing stratum should be designed within the depth range of 35 to 40 m. During construction, embedding a steel casing to at least 30 m is advised to prevent collapse of the drilled hole for the pile. It is recommended to utilize impact drills with a smaller stroke distance near the depth range where karst cavity development occurs.

## 4.6 Discussion

An MFAW detection system is developed to detect karst cavities near drilling boreholes. However, the limitations and resolution of downhole tool further inversion methods and the effects of karst cavities on piles need to be discussed as follows:

Determining the lateral distance between the borehole and the cavity relies on calculating $d = 1/2\, vt$, where $v$ and $t$ represent the velocity and arrival time of the reflected P-wave, respectively. However, accurately identifying the P-wave velocity in subsoil or bedrock using zero-offset sonic data is challenging. In field tests, the P-wave velocity can be estimated using the following approach. Firstly, the stratum type can be determined based on the amplitude of high-frequency test signals and rock core samples. When the detection depth range corresponds to bedrock, the velocity depends on lithology, RQD, and the water table. A typical estimation for weakly weathered limestone is a P-wave velocity of 2500 m/s. Additionally, the

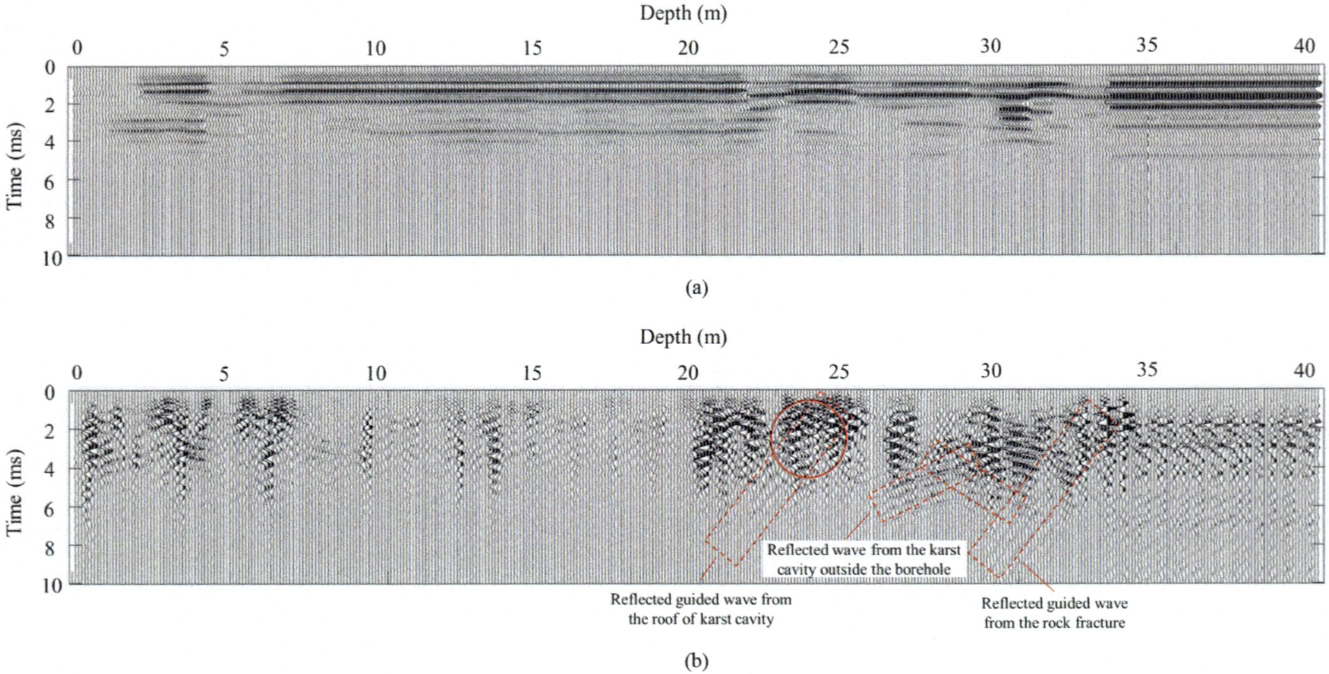


**Fig. 4.23** Signal analysis results of the low-frequency detection wave profile obtained by the no. 1 receiver whose azimuth is 37 degrees: **a** removed wave profile using median filtering, **b** median filtered wave profile

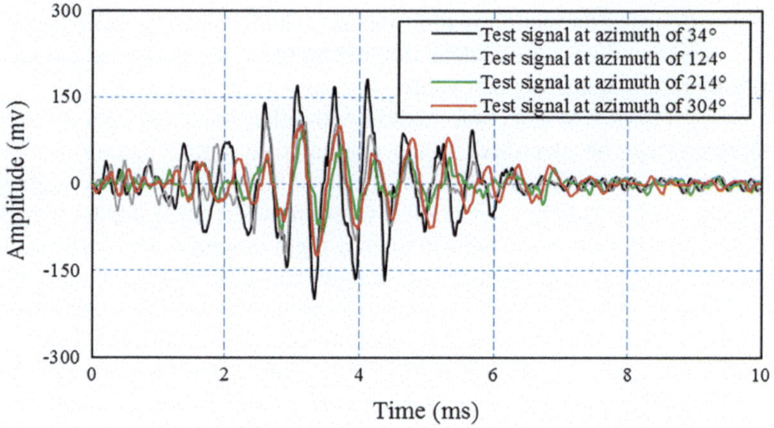


**Fig. 4.24** Median filtered signals at a depth of 27.8 m of the four different azimuths

**Fig. 4.25** Drilling core of the borehole that is in the northeast of borehole ZK-80

velocity of guided waves, which closely resembles S-wave velocity, can provide a rough estimate of the P-wave velocity. When the detection depth range is within a cavity, the velocity of the cavity infill can be determined through the following steps. If the cavity is filled with air or water, velocities of 340 m/s or 1500 m/s can be assumed, respectively. If the karst cavity contains a low-velocity solid medium, the infill material is likely to be similar to the surrounding overburden based on the origin of the karst cavity. Overburden characteristics can be estimated through the drilling of pile holes and boreholes, with results from cone penetration tests, borehole wave velocity tests, and core sample tests providing insights into the infill velocity. In this study, the sampling time length is 10 ms, and the maximum detection lateral range can be estimated using the sampling time and the velocity of the stratum. However, it should be noted that the proposed detection method and instrument have limitations in identifying multiple lateral karst voids.

Comparing the MFAW detection method with other methods, both its strengths and weaknesses can be identified. When compared to surface geophysical investigation methods like GPR, MASW, ERT, and shallow seismic tomography, the MFAW method offers superior detection resolution over a larger depth range due to its borehole-based detection and high-frequency logging source. This high-resolution detection is crucial for civil engineering designers in determining suitable bearing layers for deep foundations in karst areas. However, the MFAW method requires advanced drilling boreholes for detection, and the presence of unwanted debris sediment at the bottom of the borehole can affect the depth range of detection. In comparison to cross-hole geophysical methods such as radar and seismic computed tomography (CT), the MFAW detection method is more time and cost-efficient, aligning well with engineering progress. However, the imaging results of the MFAW method may not be as visually intuitive as those obtained from cross-hole tomography. Additionally, as a P-wave-based detection, the MFAW method may not be highly sensitive to the infill material within the karst voids. When compared to other single borehole detection methods like BARS (Li et al. 2019), the MFAW method utilizing a zero-offset acquisition system can easily detect the range beneath the borehole bottom, which is essential for determining the bearing layer. Existing BARS downhole tools with far-offset acquisition systems do not achieve this depth range. However, the FWI inversion for zero-offset borehole data may encounter local minimum problems, which can be resolved by implementing a more robust inversion strategy.

## 4.7 Conclusion

In this chapter, a signal analysis method for multi-frequency acoustic wave detection around boreholes is proposed, specifically for detecting cavities. The effectiveness of the detection system was tested through controlled indoor experiments and field applications at the Qingxi Bridge project. The following conclusions are drawn:

The integrity of rock near boreholes can be estimated using the zero-offset section of high-frequency detection, which shows a strong correlation with the energy of the high-frequency detection signal. Intact rock is indicated by a strong signal, while broken rock is characterized by alternating strong and weak signals within a short depth range. When the borehole encounters a karst cavity filled with sand, soil, or water, the test signal becomes weak. A color bar in the depth direction provides an intuitive representation of the borehole wall's integrity and offers higher precision compared to core samples.

The low-frequency acoustic detection wave field includes waves within the borehole, reflected body waves from karst cavities, and reflected guided waves from the borehole's interface. Median filtering is an effective technique for removing constant arrival time waves within the borehole, thereby enhancing the detectability of reflected waves. The integrity of the rock in the borehole, obtained through high-frequency acoustic detection, assists in identifying the reflected guided waves. After median filtering, the reflected waves from karst cavities near the borehole are enhanced.

The azimuths of karst cavities can be determined by analyzing the differential detection results obtained from azimuthal receivers. The receiver facing the cavity records higher amplitude signals, allowing for identification of the azimuthal angle using the three-dimensional electric compass in the probe.

The MFAW detection system underwent testing in a water tank and a model borehole before being applied in the field at the Qingxi Bridge project. The detection results successfully identified a karst cavity outside the borehole, contributing to the characterization of the pile bearing stratum and informing the construction method during the investigation period. The reliability of the methodology and detection system was confirmed through controlled model tests and field applications.

## References

Berenger JP (1994) A perfectly matched layer for the absorption of electromagnetic waves. J Comput Phys 114(2):185–200

Chang XM, Gao F, Lu ZT et al (2016) A study on lateral transient vibration of large diameter piles considering pile-soil interaction. Soil Dyn Earthq Eng 90:211–220

Chew WC, Liu QH (1996) Perfectly matched layers for elastodynamics: a new absorbing boundary condition. J Comput Acoust 4(04):341–359

Collino F, Tsogka C (2001) Application of the perfectly matched absorbing layer model to the linear elastodynamic problem in anisotropic heterogeneous media. Geophysics 66(1):294–307

Duncan G, Beresford G (1995) Some analyses of 2-D median fk filters. Geophysics 60(4):1157–1168

Jaganathan AP, Allouche E, Simicevic N (2010) Numerical modeling and experimental evaluation of a time domain UWB technique for soil void detection. Tunn Undergr Space Technol 25(6):652–659

Ke G, Zheng ZC (2016) Sound propagation around arrays of rigid and porous cylinders in free space and near a ground boundary. J Sound Vib 370:43–53

Lu ZT, Wang ZL, Liu DJ (2013) Study on low-strain integrity testing of pipe-pile using the elastodynamic finite integration technique. Int J Numer Anal Meth Geomech 37(5):536–550

# Chapter 5
# Inversion of Borehole Sonic Detection

## 5.1 Prologue

Full-waveform inversion (FWI) (Tarantola 1984) has emerged as a promising technique for high-resolution imaging of the subsurface. It involves iteratively minimizing an objective function that quantifies the differences between observed and simulated data. Both time-domain and frequency-domain FWI methods have demonstrated potential for accurately inverting shallow seismic data. Notably, recent real-world applications (Groos et al. 2017; Köhn et al. 2016; Tran et al. 2013; Pan et al. 2019, 2020; Wang et al. 2019; Smith et al. 2019) have showcased the feasibility of using FWI to detect underground sinkholes and near-surface structures by leveraging surface waves such as Rayleigh waves and Love waves. However, it is worth noting that most of the existing research on near-surface FWI has primarily focused on surface seismic data. As a result, there remains a gap in effective inversion strategies specifically tailored for borehole sonic data. Challenges such as local minima and cycle-skipping issues continue to hinder the development of robust and reliable FWI techniques for borehole sonic data inversion. Addressing these challenges and refining the inversion strategies for borehole sonic data are areas that require further investigation and development in future research endeavors.

This chapter introduces a novel time-domain full-waveform inversion (FWI) approach that utilizes zero-offset sonic borehole data for detecting karst voids in the subsurface. To address the challenge of local minima in zero-offset borehole data, we employ an amplitude envelope-based objective function FWI method, which provides an improved initial velocity model for correlation-based FWI. By analyzing the inverted Vs velocity profile, we can interpret modeled anomalies associated with various types of karst voids, including single voids, multiple voids, and fractures. In field tests, we compare the FWI results obtained from a single zero-offset sonic


L. Liu et al., *Detection of Karst Voids at Deep Pile Foundation*,
https://doi.org/10.1007/978-981-97-5834-0_5

borehole data with the imaging outcomes derived from four surrounding cross-hole acoustic tomography borehole profiles. This comparison showcases the capability and wide potential applications of the technique in underground construction projects.

## 5.2 Methodology of FWI

To address the challenges associated with full-waveform inversion (FWI) of zero-offset borehole data, we propose an FWI strategy that tackles two main issues. The first challenge is the absence of low-frequency data, which can lead to convergence to local minima if the initial model is far from the global minimum. Background velocity estimates derived from borehole core samples may provide some guidance, but they do not account for the significant velocity anomalies associated with karst voids and fractures. The second challenge arises from the differences in source wavelets generated at various depths. These variations can affect the accuracy of the inversion process. To overcome these challenges, our suggested FWI strategy incorporates a wave simulator that facilitates the forward modeling of synthetic data. Additionally, an adjoint simulator is employed, utilizing both envelope-based and correlation-based objective functions to calculate the model update based on the data misfit.

### *5.2.1 Forward Modeling*

We employ the Finite-Difference Time-Domain (FDTD) method to implement full-waveform inversion (FWI) in the time domain for the viscoelastic wave equation. To mitigate the numerical dispersion caused by high-frequency waves, we utilize the 8th order FDTD technique (Virieux 1986). Additionally, we incorporate a Convolutional Perfectly Matched Layer (CPML) boundary condition (Komatitsch and Martin 2007) to minimize wave reflections at the boundaries of the physical domain. Given the short distance between the transmitter and receiver (0.1 m), each individual shot can be treated as a zero-offset test. For efficient simulation of numerous zero-offset profiles, we activate multiple high-density sources that exclusively excite compressional waves simultaneously. For attenuation, Generalized Standard Linear Solid (GSLS) is used to model the propagation of P-SV waves in viscoelastic media (Robertsson et al. 1994). The stress-velocity formulation of the viscoelastic waves based on GSLS (Bohlen 2002) are shown as:

$$\rho \frac{\partial v_i}{\partial t} = \frac{\partial \sigma_{ij}}{\partial x_j} + f_i \tag{5.1}$$

$$\frac{\partial \sigma_{ij}}{\partial t} = \frac{\partial v_k}{\partial x_k}\{\pi(1+\tau^p) - 2\mu(1+\tau^s)\} + 2\frac{\partial v_i}{\partial x_j}\mu(1+\tau^s) + \sum_{l=1}^{L} r_{ijl} \ \mathit{if}\ i = j \tag{5.2}$$

$$\frac{\partial \sigma_{ij}}{\partial t} = \left(\frac{\partial v_k}{\partial x_k} + \frac{\partial v_k}{\partial x_k}\right)\mu(1+\tau^s) + \sum_{l=1}^{L} r_{ijl} \quad \mathit{if}\ i \neq j \tag{5.3}$$

where $\sigma_{ij}$ denotes the *ij*th component of the stress tensor, $v_i$ are the components of the particle velocities. $f_i$ are the components of the external body force. The levels of attenuation for P-waves and S-waves are represented by $\tau^p$ and $\tau^s$. These attenuation levels are associated with the seismic quality factors $Q_p$ and $Q_s$, respectively:

$$\tau^p = 2/Q_p \tag{5.4}$$

$$\tau^s = 2/Q_s \tag{5.5}$$

In our simulation, we assign distinct seismic quality factors, $Q_s$ (S-wave) and $Q_p$ (P-wave), to the grid points to capture the spatial distribution of attenuation. Notably, regions such as boreholes, karst voids, or fractures exhibit lower values of $Q_s$ and $Q_p$, resulting in substantial amplitude decay and velocity dispersion.

### 5.2.2 *Inversion Strategy*

To perform FWI for the detection of karst voids using zero-offset sonic borehole data, we employ envelope-based and correlation-based objective functions, dividing the process into the following key steps: (1) Establish an appropriate initial model, typically estimated from borehole rock samples in field processing scenarios. (2) Determine the source wavelet. In field data inversion, the source wavelet is estimated through controlled tests in a water tank. In synthetic data processing, the true source wavelet is directly utilized. (3) Apply a low-pass filter to both the observed data and modeled data. (4) Perform envelope-based objective function FWI. This involves calculating the forward wavefield, residual, gradient, and step length based on the envelope objective function. The model parameters are updated accordingly. (5) Execute correlation-based objective function FWI within the same frequency range. The updated model is loaded, and the FWI process is repeated based on the correlation-based objective function. (6) Gradually increase the high-cut frequency of the filter and repeat steps (4) and (5) for each frequency range, progressing from low-frequency to high-frequency inversion (Fig. 5.1).

To improve the initial model for subsequent inversions, we initially employ the envelope objective function-based FWI (Bozdag et al. 2011). In the traditional FWI approach, the estimation of similarity between the modeled and observed data relies on an L2 norm difference objective function.

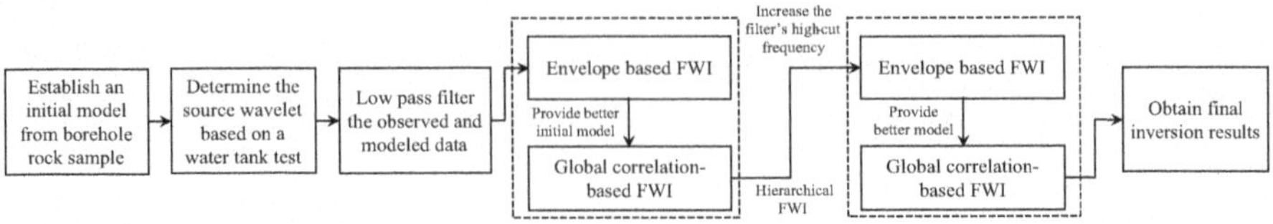


**Fig. 5.1** Flow chart of FWI employing the envelope-based and correlation-based objective functions

$$E = \frac{1}{2}\sum_{s,r}\int_{0}^{T}[u(x_s, x_r, t) - (x_s, x_r, t)]^2 dt \tag{5.6}$$

where $u(x_s, x_r, t)$ is the modeled data and $u_0(x_s, x_r, t)$ is the observed data at the given source $x_s$ and receiver $x_r$ locations. The locations of the source and receiver are assumed to be the same since the observed data have a zero-offset profile (i.e., $x_s = x_r$). To evaluate the dissimilarity between the envelopes of the modeled and observed data, we employ the envelope-based objective function described in Eq. 5.7. This objective function quantifies the misfit between the envelope characteristics of the modeled and observed data.

$$E = \frac{1}{2}\sum_{s,r}\int_{0}^{T}[e(x_s, x_r, t) - e_0(x_s, x_r, t)]^2 dt = \frac{1}{2}\sum_{s,r}\int_{0}^{T} K^2 dt \tag{5.7}$$

where $e(t)$ is the envelope of $u(t)$. The envelope is calculated by the Hilbert transform (Taner et al. 1979).

$$e(t) = \sqrt{u^2(t) + v^2(t)} \tag{5.8}$$

$$v(t) = H(u(t)) = \frac{1}{\pi} p.v. \int_{-\infty}^{+\infty} \frac{u(\tau)}{t - \tau} d\tau \tag{5.9}$$

where $v(t)$ is the imaginary component of $u(t)$ and is obtained by Eq. 5.9 using the Cauchy principal value (denoted here by $p.v.$).

By applying the Hilbert transform to the original test signal, low-frequency components are generated. This allows for the extraction of the envelope signal, which contains valuable low-frequency information suitable for estimating the long wavelength velocity model. The gradient of the envelope objective function [Wu et al. 2014] is expressed as follows:

$$\frac{\partial E}{\partial m} = \sum_{s,r} \int_0^T K \cdot e^{-1}(x_s, x_r, t) \left[ u(x_s, x_r, t) \frac{\partial u(x_s, x_r, t)}{\partial m} + v(x_s, x_r, t) \frac{\partial v(x_s, x_r, t)}{\partial m} \right] dt \tag{5.10}$$

To perform full-waveform inversion using the velocity model obtained from the envelope-based objective function inversion, we utilize a global correlation-based objective function (Choi and Alkhalifah 2012). This objective function, which takes into account the global correlation, is expressed as follows:

$$\begin{aligned} E &= \sum_{s,r} \int_0^T \left[ -\frac{u(x_s, x_r, t)}{|u(x_s, x_r, t)|} \cdot \frac{u_0(x_s, x_r, t)}{|u_0(x_s, x_r, t)|} \right] dt \\ &= \sum_{s,r} \int_0^T [-\hat{u}(x_s, x_r, t) \cdot \hat{u}_0(x_s, x_r, t)] dt \end{aligned} \tag{5.11}$$

The global correlation of two signals is equal to one when the modeled and observed signals are identical and less than one when the similarity is weak. Correlation-based FWI controls the amplitude of modeled data based on the similarity between the observed and modeled data. Thus, correlation-based FWI can mitigate the error caused by the differences of source wavelets excited at different depths and with inaccurate acquisition geometries. The gradient of global correlation-based function (Choi and Alkhalifah 2012) is shown as:

$$\frac{\partial E}{\partial m} = \sum_{s,r} \int_0^T \left[ \begin{array}{c} \frac{\partial u(x_s, x_r, t)}{\partial m} \cdot \frac{1}{|u(x_s, x_r, t)|} \\ \{\hat{u}(x_s, x_r, t)(\hat{u}(x_s, x_r, t) \cdot \hat{u}_0(x_s, x_r, t)) - \hat{u}_0(x_s, x_r, t)\} \end{array} \right] dt \tag{5.12}$$

A preconditioning operator is used to approximate the inverse of the Hessian matrix. The operator uses a zero-lag correlation of the absolute value of the forward wavefield with an approximation of the receiver Greens function contribution:

$$H_a^{-1} = \left\{ \int_0^T dt |u(x_s, x_r, t)|^2 \left[ \operatorname{arcsin} h\left( \frac{x_r^{\max} - x}{z} \right) - \operatorname{arcsin} h\left( \frac{x_r^{\min} - x}{z} \right) \right] \right\}^{-1} \tag{5.13}$$

where $x_r^{\min}$, $x_r^{\max}$, and $x_s$ are the minimum, maximum receiver and source positions, respectively.

We utilize the quasi-Newton limited memory Broyden-Fletcher-Goldfarb-Shanno (L-BFGS) method (Köhn et al. 2016) to address the nonlinear optimization problem in FWI. While both P-wave and S-wave velocities are updated during the inversion process, our primary focus is on identifying voids based on the inverted S-wave velocity profile. This is due to the distinct S-wave velocity of karst voids or fractures, which differs from the background S-wave velocity of intact limestone. Moreover, the gradient exhibits relatively higher resolution (Wang et al. 2019) since the wavelength of the S-wave is typically shorter compared to that of the P-wave.

## 5.3 Numerical Study

Various types of karst voids can be observed, including single cavities, multiple cavities, and dissolution-enhanced fractures. We have designed numerical examples that represent these three karst features. Borehole rock samples and well logs obtained during drilling provide information about the subsurface lithology, which is used to estimate the background velocity and density. The S-wave velocities ($V_s$) of the three true models are depicted in Fig. 5.2a, b, and c. The background $V_s$ ranges from 1250 to 1600 km/s. The corresponding background P-wave ($V_p$) and density ($\rho$) vary from 3450 to 4100 km/s and 1.7 to 2.2 g/cm$^3$, respectively. Seismic qualities ($Q_s$) and ($Q_p$) range from 105 to 70, with values increasing with formation depth. The geometries of the true P-wave and density models mirror those of the S-wave models. A vertical borehole with a diameter of 0.15 m is centered at x = 1.15 m, characterized by $V_s = 0$, $V_p = 1500$ m/s, $\rho = 1000$ kg/m$^3$, and $Q_s =$ Qp $= 30$. In the true single karst model (Fig. 5.3a), there is a round-shaped water-filled void with a diameter of 2 m ($V_s = 0$, $V_p = 1500$ m/s, ρ = 1000 kg/m3, and $Q_s = Q_p = 30$). It is centered at $x = 4$ m and $y = 10$ m. In the true multiple karst void model (Fig. 5.2b), there are two parallelogram-shaped low-velocity anomalies with different dip directions. They are centered at $x = 4.5$ m, $y = 7.5$ m, and $x = 3.5$ m, $y = 15$ m. The upper void ($V_s = 600$ m/s, $V_p = 2200$ m/s, $\rho = 1400$ kg/m$^3$, and $Q_s = Q_p = 40$) has a wider width of 0.5 m compared to the lower void ($V_s = 0$, $V_p = 1500$ m/s, $\rho = 1000$ kg/m$^3$, and $Q_s = Q_p = 30$). The presence of finite S-wave velocity in the upper void indicates the presence of clay or sand infill. In the true rock fracture model (Fig. 5.2c), a low-velocity anomaly with a width of 0.5 m ($V_s = 0$, $V_p = 1500$ m/s, $\rho = 1000$ kg/m$^3$, and $Q_s = Q_p = 30$) can be observed between depths of 5 and 10 m. To emphasize the influence of velocity anomalies, the initial model (Fig. 5.2d) has been created to be an exact replicate of the true velocity model, excluding the anomalies.

In the synthetic case studies, the model dimensions are set at 20 m in depth and 7.5 m in width. To discretize the model, a spatial interval of 0.025 m is used in both the $x$- and $y$-directions, resulting in a grid size of 800 nodes in the x-direction and 300 nodes in the $y$-direction. To effectively handle wave propagation at the boundaries, Convolutional Perfectly Matched Layers (CPMLs) are implemented. For the source waveform, a Ricker wavelet with a dominant frequency of 2 kHz is employed. The time sampling interval is set to 0.002, and the duration of the simulation window is 6 ms. To solve the viscoelastic wave equation, the 8th order Finite-Difference Time-Domain (FDTD) method is utilized.

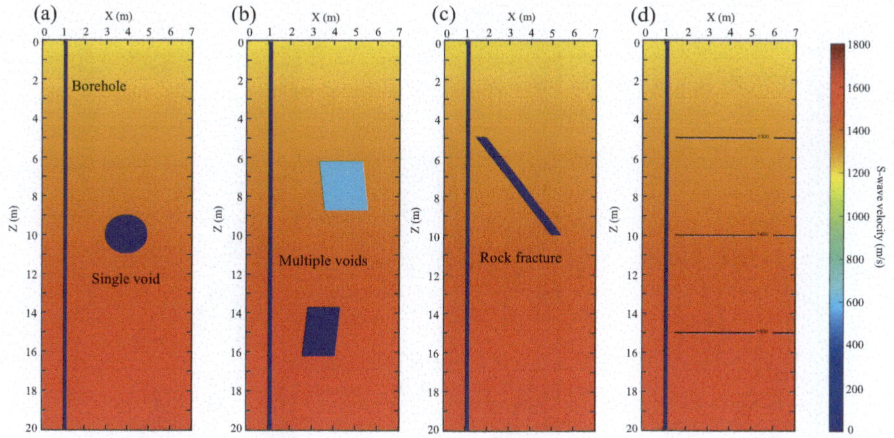


**Fig. 5.2** S-wave velocity models. **a** Single void. **b** Multiple voids. **c** Rock fracture. **d** Initial model

The synthetic zero-offset data for the horizontal vx component, as shown in Fig. 5.3a, b, and c, are generated from the true model. The simulated survey consists of 200 channels with receivers placed at 0.1 m intervals along the borehole. The source line is aligned with the receiver locations, and it generates horizontal compressional waves simultaneously. These waves result in body-wave (P and SV) diffractions from the front and back sides of voids, which are clearly observed in the horizontal records depicted in Fig. 5.3a and b. In the synthetic data, multiple reflected waves within the borehole are observed, and they overlap in the time window due to reflections from the nearby void (Fig. 5.3c). It is important to note that, in practical scenarios, it is challenging to completely eliminate the influence of the borehole, as the time duration of the source wavelet is often much longer than the ideal Ricker wavelet used in the simulation. Consequently, the multiple reflections within the borehole prolong the effective reflection time window from the void.

The amplitude envelope of the original synthetic zero-offset data is obtained through the application of the Hilbert transform, as depicted in Fig. 5.4a, b, and c. For instance, let's consider the record from a 10 m depth in the true single void model, shown in Fig. 5.5a and b. It is observed that the original signal exhibits a dominant frequency of 3 kHz, with a frequency range of 2 to 4 kHz. However, upon examining the envelope of the signal, it is found that the dominant frequency is reduced to less than 1 kHz, and the frequency range is limited to less than 2 kHz. This observation indicates that the nonlinear transform employed in obtaining the envelope signal has the capability to introduce low-frequency information into the spectrum, even when the original waveform data lacks such low-frequency components. This inclusion of low-frequency components in the envelope signal proves beneficial in addressing issues related to local minimum convergence and cycle-skipping (Chi et al. 2014), as discussed in previous studies. By incorporating the low-frequency information, these problems can be mitigated and enhance the effectiveness of the inversion process.

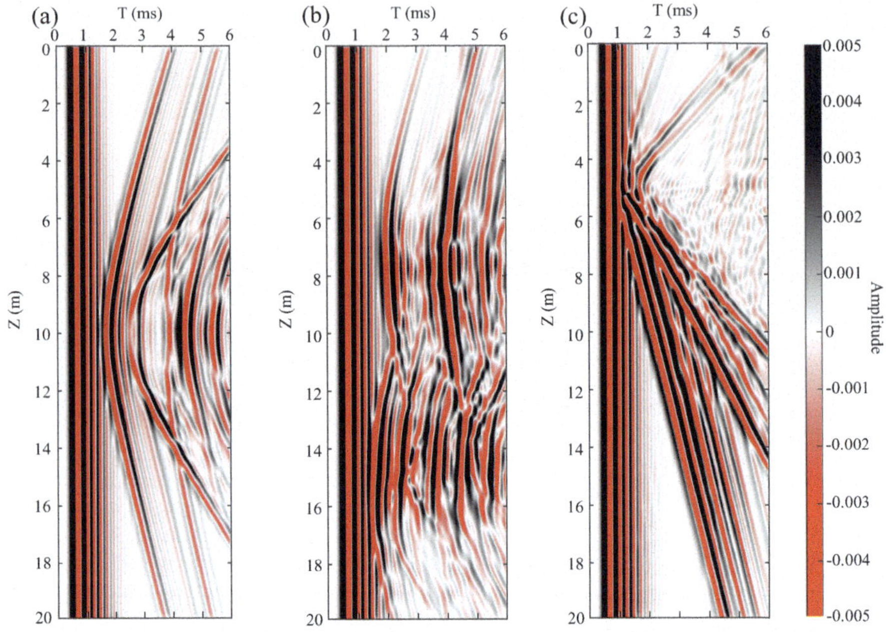


**Fig. 5.3** Synthetic zero-offset data of the horizontal $v_x$ component. **a** Single void. **b** Multiple voids. **c** Rock fracture

The inversion process for each scenario (depicted in Figs. 5.6, 5.7, and 5.8) can be divided into four stages, progressing from the low-frequency range (~ 0.3 kHz) to the high-frequency range (~ 1.5 kHz) [Bunks et al. 1995; Kurzman 2012]. This is achieved through the alternating use of envelope- and correlation-based objective functions. Prior to each iteration, both the synthetic and observed data undergo a low-pass filtering procedure. The high-cut frequency gradually increases during the inversion process, with the inversion stopping either when the high-cut frequency reaches a certain threshold or when the difference in objective function misfit between the current and previous iterations falls below 1%. To demonstrate the effectiveness of the dual envelope and correlation FWI strategy, two typical frequency range inversion results are presented. In the case of the single void scenario (Fig. 5.6), the envelope-based inversion of the Vs profile in the low-frequency range (Fig. 5.6a) reveals a rough depiction of a low-velocity circular anomaly with a midpoint depth of around 10 m. However, the exact shape and size of the anomaly are not clearly discernible. Subsequently, the correlation-based FWI (Fig. 5.6b) further refines the image of the anomaly. Moving on to the high-frequency range, the inverted Vs profiles (Fig. 5.6c and d) exhibit a nearly zero circular-shaped S-wave velocity anomaly, which closely matches the true model. In this simple model, the difference between the envelope-based and correlation-based FWI results is not significant. Although there is a high-velocity artifact surrounding the low-velocity cavity, it does not hinder

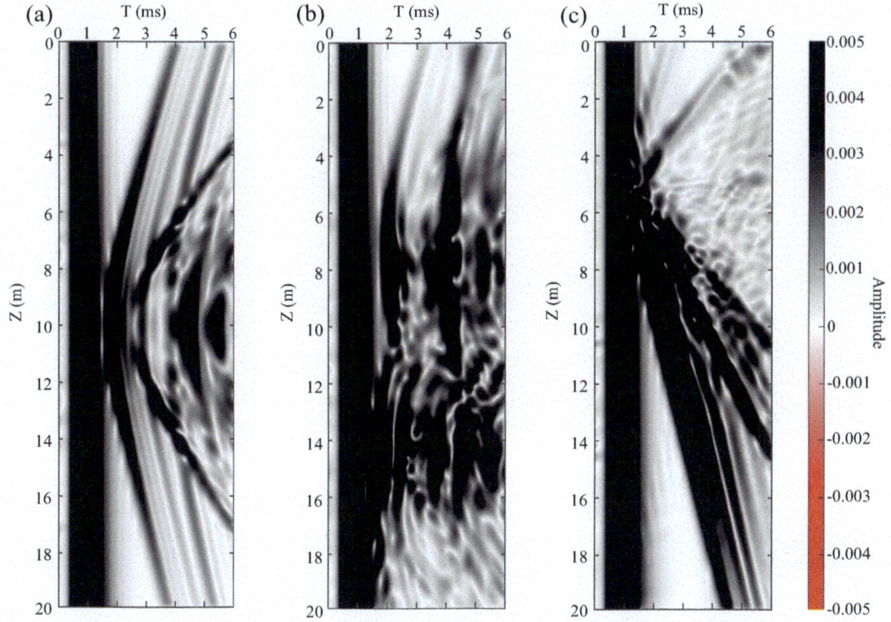


**Fig. 5.4** Envelope profile of synthetic zero-offset data of the horizontal component. **a** Single void. **b** Multiple voids. **c** Rock fracture

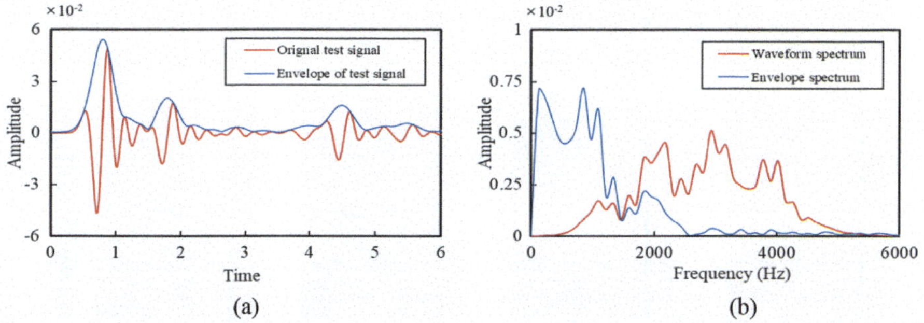


**Fig. 5.5** A shot record for a single void model and its envelope. **a** Shot record and its envelope in the time domain. **b** Shot record and its envelope in the frequency domain

the identification of the karst void, as the void consistently appears as a low-velocity anomaly.

Figure 5.7 illustrates the inverted models depicting multiple karst voids. In the low-frequency range, the envelope-based inversion of the Vs profile reveals a discernible low-velocity anomaly at a depth of 15 m. However, the presence of a higher velocity void at a depth of 7 m is hardly noticeable, as shown in Fig. 5.7a. The correlation-based FWI improves the visibility of the upper void, as depicted in Fig. 5.7b. Both

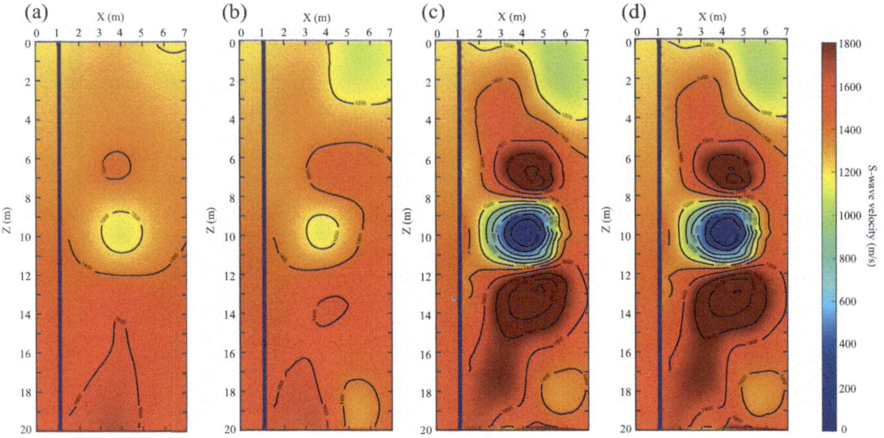


**Fig. 5.6** Full-waveform inversion for a single void scenario. **a** Stage 1: Envelope-based FWI in the low-frequency range. **b** Stage 2: Correlation-based FWI in the low-frequency range. **c** Stage 3: Envelope-based FWI in the high-frequency range. **d** Stage 4: Correlation-based FWI in the high-frequency range

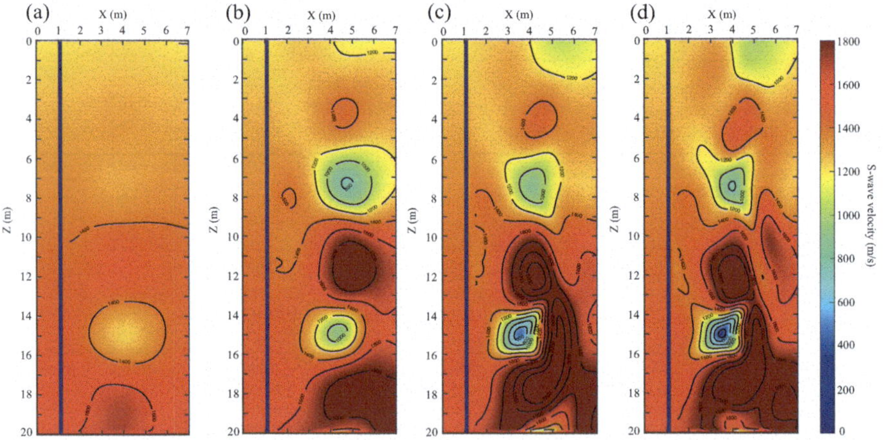


**Fig. 5.7** Full-waveform inversion for the multiple void scenario. **a** Stage 1: Envelope-based FWI in the low-frequency range. **b** Stage 2: Correlation-based FWI in the low-frequency range. **c** Stage 3: Envelope-based FWI in the high-frequency range. **d** Stage 4: Correlation-based FWI in the high-frequency range

inverted void images appear as round-shaped voids. In the high-frequency range, the envelope-based FWI successfully captures two parallelogram-shaped low-velocity anomalies with different dip directions (Fig. 5.7c). However, the velocities of the voids in the inverted S-wave profile still significantly differ from those in the true

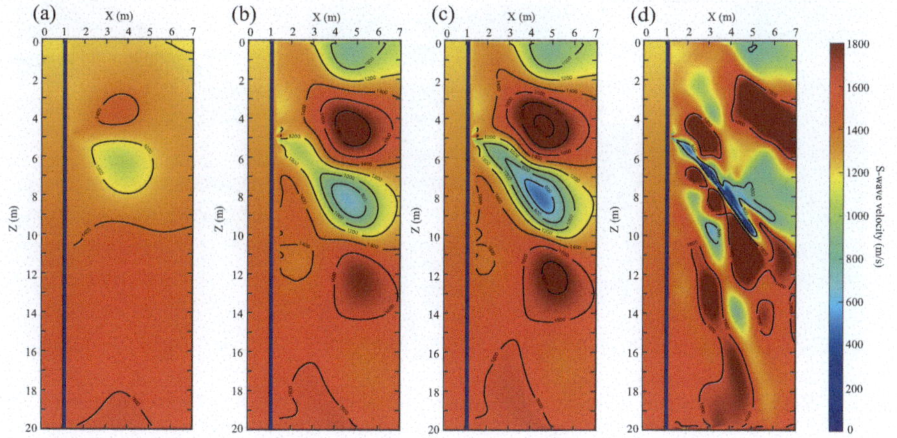


**Fig. 5.8** Full-waveform inversion for the rock fracture scenario. **a** Stage 1: Envelope-based FWI in the low-frequency range. **b** Stage 2: Correlation-based FWI in the low-frequency range. **c** Stage 3: Envelope-based FWI in the high-frequency range. **d** Stage 4: Correlation-based FWI in the high-frequency range

model (Fig. 5.2b). The correlation-based FWI enhances the estimation of the velocities within the voids and improves their shape definition (Fig. 5.7d). It is important to note that the presence of multiple voids negatively impacts the inversion process, making it more challenging to accurately estimate the velocity and shape due to the complex overlapping of backscatter energy from the multiple voids.

In the inversion process for the rock fracture scenario, the envelope-based FWI in the low-frequency range provides an improved initial model (Fig. 5.8a) for subsequent inversions. This initial model exhibits a round-shaped low-velocity anomaly that corresponds to the location of the fracture. The correlation-based FWI, focusing on the S-wave profile, presents a low-velocity zone with a linear fracture near the borehole and a round shape farther away from the borehole (Fig. 5.8b). In the high-frequency range, the envelope-based FWI refines the velocity estimation within the fracture void, bringing it closer to that in the true model (Fig. 5.8c). The correlation-based inversion of the S-wave velocity profile (Fig. 5.8d) suggests the presence of a 0.5 m-wide low-velocity fracture void from the depths of 5 m to 10 m.

## 5.4 Field Test

A field test was conducted on a freeway extension project called Guang-Qing, which involves the replacement of a four-lane road with a new eight-lane freeway. The project spans Guangzhou City and Qingyuan City in China and has a total length of 57.6 km. A significant portion of 33 km is situated in a karst area, as depicted by the red line in Fig. 5.9a. This karst area encompasses over 10 bridges and 3000 piles.

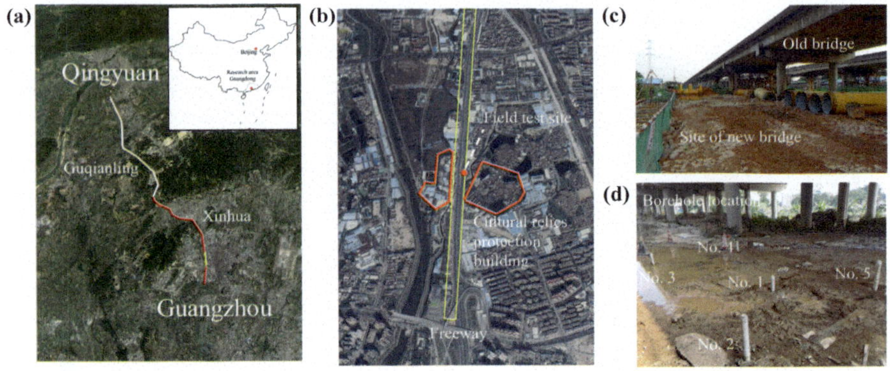


**Fig. 5.9** Regional profile of the Guang-Qing freeway extension project. **a** Overview map of the project. **b** Overview map of the field test. **c** Schematic of the freeway extension. A new bridge will be built next to the old bridge. **d** Schematic of borehole locations

One of the bridges within the karst area is illustrated in Fig. 5.9 b, and it serves as the location for the field test. Due to the detrimental effects of karst voids on pile bearing capacity and the safety of nearby structures, borehole sonic detection is employed in this project to identify the presence of karst voids in close proximity to the pile locations.

## *5.4.1 Field Geology and Data Acquisition*

Zero-offset sonic borehole data were collected at the pile locations designated for the construction of the new bridge (see Fig. 5.10c and d). The site was equipped with a set of 5 open boreholes, each having a diameter of 12.5 cm and extending to a depth of 45 m. These boreholes were arranged in a square pattern, with one central borehole (as illustrated in Fig. 5.9). The geological composition of the overburden in the study area primarily comprises upper Pleistocene and Holocene materials, with an approximate thickness of 20 m. These materials consist of marine clay, loam, sand, gravel, and silt. The underlying bedrock consists of soluble limestone belonging to the Carboniferous Huantian Group and the Shizizi Formation [50]. Karst features commonly observed in this area include cavities, cutters, and pinnacles. The distribution of these karst voids is highly irregular, as evidenced by the rock samples obtained from the boreholes (shown in Fig. 5.10). Boreholes No 1, 2, 3, and 5 encounter multiple karst voids, while borehole No. 4 borehole rock sample is intact. It should be noted that the infill material within the karst voids varies, ranging from clay and silt to water or air. Precisely identifying the nature of this infill material solely through borehole drilling can be challenging.

A single borehole sonic test was performed in the No. 1 borehole within the depth range of 20 m to 45 m. The probe was carefully lowered from the top to

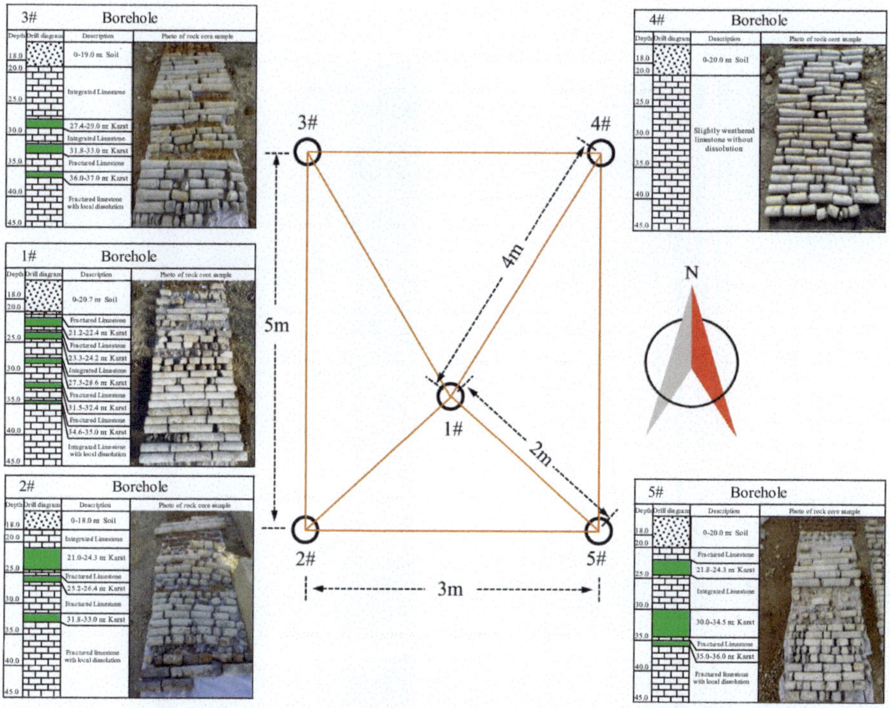


**Fig. 5.10** Schematic diagram of boreholes tested in the field

the bottom of the borehole, transmitting signals at regular intervals of 0.1 m. To enhance the quality of the collected data, various conventional signal processing techniques were applied. These included zero-time correction, bandpass frequency filtering (10–7 kHz), gain control, normalization, and other standard methods. The processed zero-offset profile and the corresponding envelope spectrum within the depth range of 22.5 to 42.5 m (consisting of 200 traces) are displayed in Fig. 5.11. Despite the application of these processing techniques, the signal-to-noise ratio (S/N ratio) of the backscatter from the voids remains relatively low. This is primarily due to the presence of strong multiple reflected waves in the borehole that tend to obscure the backscatter originating from the voids. While median filters can be utilized to remove vertical events such as waves within the borehole, it is important to note that these filters may also inadvertently eliminate the valuable backscatter signals from voids or fractures that run parallel to the borehole.

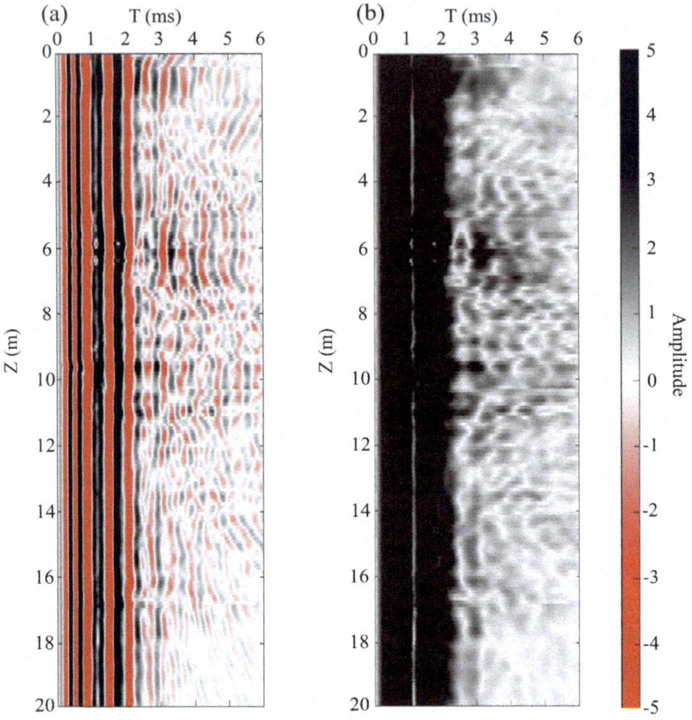


**Fig. 5.11** **a** Processed zero-offset data from a depth of 22.5 to 42.5 m (200 traces in a depth range of 20 m. To normalize the depth, we set the depth of 22.5 m as the zero of Z) in borehole No. 1 and **b** corresponding envelope data

## *5.4.2 Inversion Results*

To facilitate the inversion process of the zero-offset sonic borehole data, we conducted a measurement of the source wavelet in a water tank using the identical downhole tool employed in the field test. The water tank provided a suitable environment for this measurement, as it lacks any borehole and thus eliminates the occurrence of reflections that could interfere with the data. The estimated source wavelet, as shown in Fig. 5.12 exhibits a dominant frequency of approximately 1.7 kHz. However, it is important to note that the estimated wavelet displays more sidelobes compared to the ideal Ricker wavelet. This increased presence of sidelobes introduces significant interference, making it challenging to accurately identify the effective backscatter originating from the voids within the data.

A 1D preliminary model (Fig. 5.13a) is derived from the borehole core samples, serving as the starting point for the inversion process. The elastic parameter in this model is smoothed and varies solely with depth. The dimensions of the model are 20 m in depth and 5.25 m in width, with a spatial interval of 0.025 m. To define the sections of the model that should be updated during the inversion, a horizontal taper

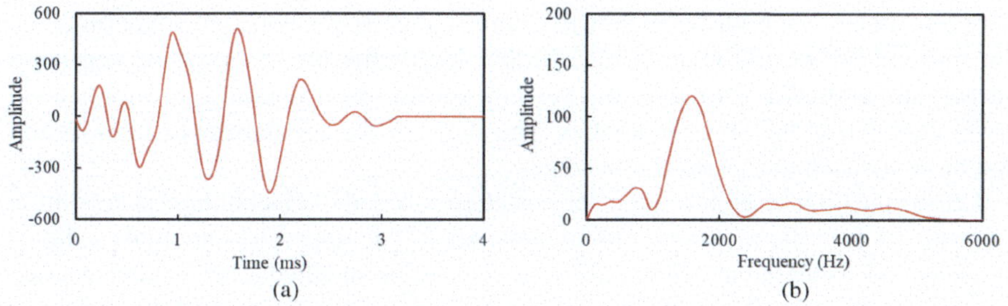


**Fig. 5.12** **a** Estimated wavelet based on controlled water tank test and **b** corresponding spectrum

is employed on the gradient. To initiate the inversion, the envelope of the processed zero-offset data (Fig. 5.11b) is utilized to generate a more refined initial model. This envelope-based approach aids in achieving improved accuracy and reliability in subsequent correlation-based FWI iterations.

The first stage of FWI inversion produces an inverted Vs profile (Fig. 5.13b) revealing three anomalies located at depths ranging from 0 to 2 m, 5.5 to 7 m, and 9.5 to 10 m. These anomalies exhibit lower Vs values compared to the background shear wave velocity but remain considerably higher than the velocities associated with clay-filled or water-filled karst voids. However, the inverted profile does not accurately capture the slope of the three anomalies. The application of correlation-based FWI yields an improved Vs profile (Fig. 5.13c) with enhanced resolution. It identifies an upward low-velocity anomaly at a depth of 16 m. The upper anomaly displays horizontal development, while the two lower anomalies slope downward.

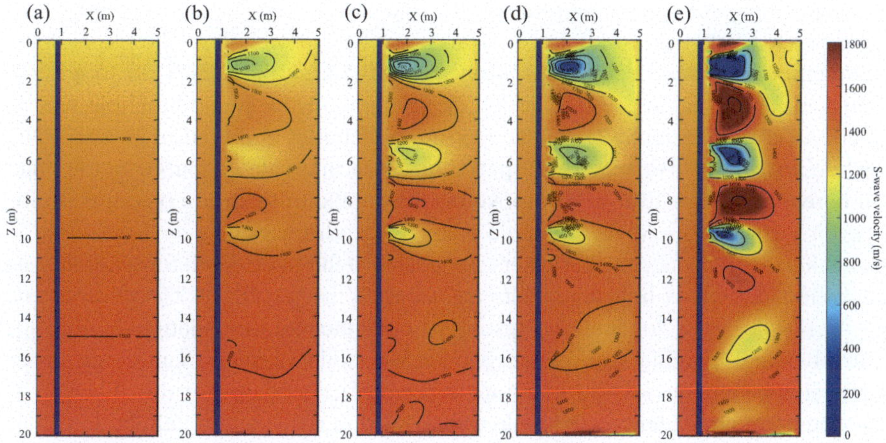


**Fig. 5.13** Full-waveform inversion for field data. **a** Initial model. **b** Stage 1: Envelope-based FWI in the low-frequency range. **c** Stage 2: Correlation-based FWI in the low-frequency range. **d** Stage 3: Envelope-based FWI in the high-frequency range. **e** Stage 4: Correlation-based FWI in the high-frequency range

Subsequently, in the high-frequency inversion using the envelope objective function, the inverted Vs profile (Fig. 5.13d) further highlights the velocity discrepancies among the anomalies. Notably, the velocity of the upper anomaly is significantly lower than that of the other anomalies. The shape of the anomalies becomes clearer and more focused in this inversion stage.

During the final inversion stage, the correlation-based FWI technique successfully reconstructs four distinct low-velocity anomalies. The uppermost anomaly exhibits an elongated rectangular shape, indicating a water or air-filled karst void measuring approximately 2 m × 2 m. Its S-wave velocity is notably low, around 10 m/s. Adjacent to the first void, a second rectangular anomaly is observed, enveloping a downward low-velocity void. The shear wave velocity within this void is not as low as that of the first void, suggesting the presence of clay or sand infill at the bottom and water infill in the upper portion. At a depth of 10 m, a smaller downward void forms the third anomaly. Similar to the second void, its infill comprises clay or sand. The lowest low-velocity anomaly is likely attributed to a weathered limestone or soft interlayer that extends upward for approximately 4 m. The S-wave velocity within this anomaly shows a relatively higher value, ranging from 16 to 14 m in depth. Overall, the envelope and correlation-based FWI strategy successfully captures the locations and shapes of potential karst voids near the borehole. Moreover, it enables the identification of shear velocity differences within these karst voids.

### *5.4.3 Cross-Hole Tomography Validation*

To verify the reliability of the FWI method, a comparison is made between the FWI imaging results and cross-hole tomography data. Four cross-hole tomography tests are conducted between borehole No. 1 and four other boreholes. The tomography investigation covers a depth range from the bedrock surface (21.5 m) to a depth of 43.5 m, encompassing the detection depth range of the single borehole test. In the cross-hole tomography tests, the acoustic transmitter is placed in borehole No. 1, while a string of five receivers, spaced at 0.5 m intervals, is placed in the other four boreholes. Water filling is employed in the boreholes to ensure proper coupling between the sensors and the borehole walls. A depth-measuring device is used to ensure that the transducers and sensors are located at the same depth for accurate data acquisition. The cross-hole sonic data are acquired using a JL-UCT(A) instrument, specifically designed for integrity testing of concrete deep foundations. Given the relatively small imaging section between the boreholes (2 m and 4 m in this case), higher signal frequencies (20 kHz) are utilized. The receivers of the JL-UCT(A) instrument are equivalent to the hydrophone used in the single borehole downhole tool (Liu et al. 2020) (Fig. 5.14).

The first arrival traveling time data from the cross-hole tests are inverted using the simultaneous iterative reconstruction technique (SIRT) and shortest path method. The resulting P-wave velocity tomograms (Fig. 5.15a–d) are found to be consistent with the borehole rock samples (Fig. 5.10). The interpretation of the tomograms reveals

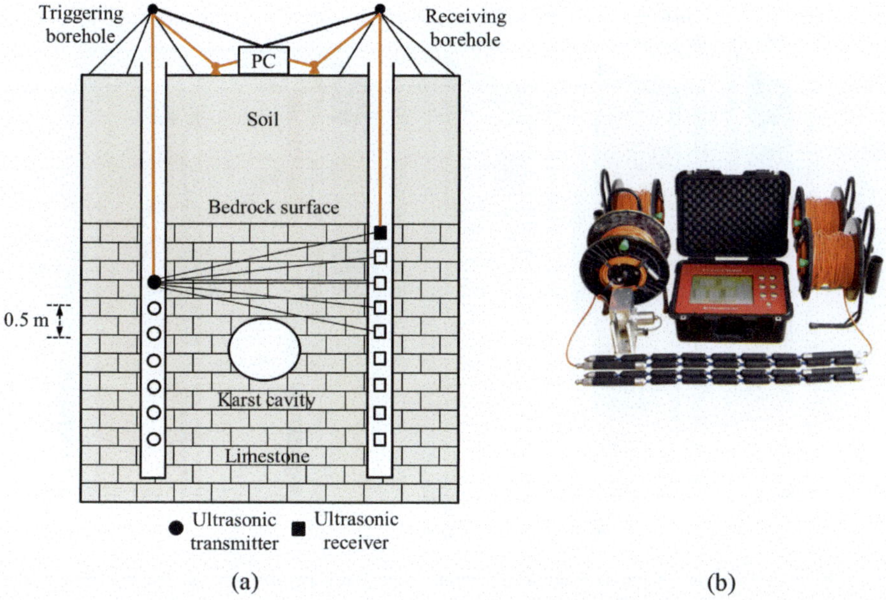


**Fig. 5.14** **a** Schematic diagram of cross-hole sonic tomography. **b** Field ultrasonic acquisition system

that karst voids without any infill correspond to P-wave velocities of less than 0.4 km/s. Areas with light blue shading, exhibiting P-wave velocities between 0.4–1.0 km/s, are identified as cohesive soil or sand. P-wave velocities of approximately 1.5 km/s suggest the presence of water or mixed infill. Green zones with P-wave velocities ranging from 1.5–2.5 km/s are interpreted as fractured or local karst limestone. To assess the agreement between the inverted P-wave velocity profiles and the low S-wave velocity anomalies obtained from FWI, comparisons are made regarding the shapes, sizes, positions, and infill characteristics of the voids in the profiles (Fig. 5.15a-d) and the S-wave velocity anomalies (Fig. 5.15e) inverted by FWI.

The low S-wave velocity anomalies obtained from FWI can be correlated with the voids observed in the cross-hole P-wave velocity tomograms, irrespective of the direction. The upper anomaly in Fig. 5.15e, with a S-wave velocity of zero, corresponds to the air-filled voids seen in Fig. 5.15d and e at depths ranging from 0 to 2 m. The second rectangular anomaly and the third downward voids, with clay and water infill in the FWI results (Fig. 5.15e), can also be identified as the voids located at depths of 6 and 10 m in the cross-hole tomograms (Fig. 5.15b and d). The lowest anomaly observed in both the FWI results and the tomograms between boreholes No. 1 and No. 3 (Fig. 5.15b) represents a fracture or a localized zone of karst limestone. Overall, there is consistency between the shapes, depths, and infill characteristics of the voids observed in the FWI S-wave velocity inversion result and those in the cross-hole P-wave velocity tomograms.

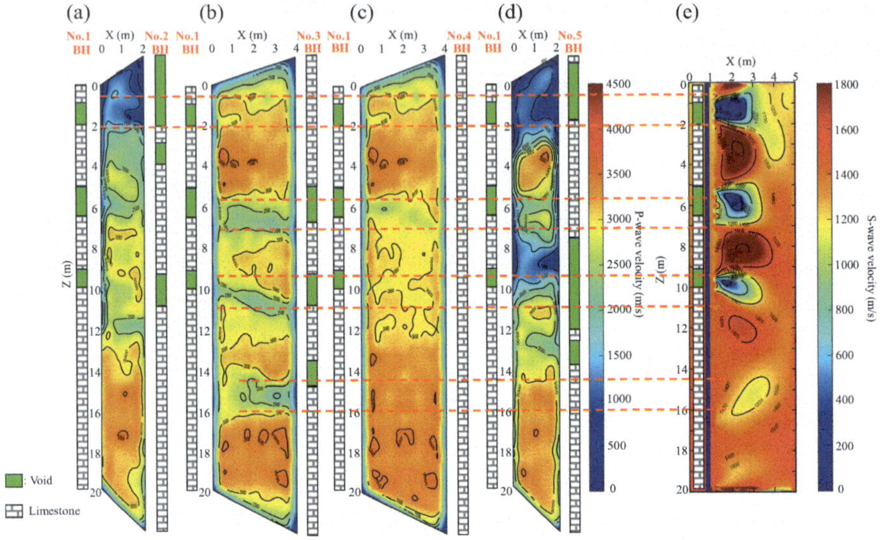


**Fig. 5.15** Comparison of cross-hole tomography and FWI results for single borehole data. **a** Tomography between boreholes No. 1 and 2. **b** Tomography between boreholes No. 1 and 3. **c** Tomography between boreholes No. 1 and 4. **d** Tomography between boreholes No. 1 and 5. **e** Final FWI result for borehole No. 1 data. Borehole core logs identifying void and limestone sections are displayed at the corresponding well locations

## 5.5 Discussion

The analysis of zero-offset borehole sonic data can provide valuable insights for transportation, infrastructure, and mining construction projects. The objective of this study was to develop a FWI imaging strategy using single borehole zero-offset data for the detection of karst voids. Several aspects of the proposed FWI strategy using zero-offset data warrant further discussion.

The combination of envelope- and correlation-based objective functions helps mitigate the local minimum problem. Due to the heterogeneous properties of karst, the initial model for void detection is often far from the exact model. The envelope signal captures the overall features of the borehole data, and the addition of low-frequency information into the spectra prevents the model from falling into a local minimum. The envelope FWI provides a better velocity model in each frequency stage and offers initial indications of low-velocity anomalies. The correlation FWI further improves the inversion based on the velocity model provided by the envelope FWI.

The source wavelet can be estimated through controlled water tank tests. For zero-offset borehole data, the wavelet is triggered in the water-filled borehole, and its shape remains relatively unchanged until it is recorded by the receiver due to the small interval between the transmitter and receiver.

However, the current FWI strategy for zero-offset borehole data still encounters challenges in the high-frequency inversion stage. The dominant frequencies of the source wavelet in numerical simulations and field tests are 2 kHz and 1.7 kHz, respectively. The FWI strategy involves low-pass filtering of the observed data before each iteration. The final inverted results presented in this study are obtained by setting the high-cut frequency to approximately 1.5 kHz to mitigate the local minimum problem. Inversion with a higher-cut frequency does not significantly update the model parameters. To improve accuracy, a better acquisition system with more ray path coverage is required to address near-surface engineering problems and enhance the robustness of the FWI strategy.

The FWI of zero-offset borehole data lacks azimuth information. The azimuths of anomalies are typically analyzed by comparing the amplitude energy of the backscatter recorded by azimuthal receivers. However, the backscatter amplitude only impacts the inverted velocity models in full-waveform inversion, making identification of energy differences complex. Azimuthal FWI for borehole data will be explored in future studies. In the bridge extension project, detecting any karst anomaly at the pile location indicates the need to deepen the borehole to a competent foundation bearing layer.

The comparison between FWI and cross-hole tomography in this study uses different velocity components. The FWI results are presented using S-wave velocity, while cross-hole tomography is based on P-wave velocity. The infill, geometry, and location of karst voids inverted by FWI can only be roughly verified by cross-hole tomography. Future studies will consider incorporating S-wave sensors in the borehole or developing an effective extraction algorithm for S-wave velocity tomography. Additionally, more accurate full-waveform inversion of P-wave velocity is necessary for comprehensive interpretation of zero-offset sonic borehole data.

## 5.6 Conclusions

In this chapter, a novel approach combining envelope and correlation objective functions is presented for full-waveform inversion (FWI) of zero-offset sonic borehole data to detect karst voids at pile foundations. The key findings of this study are as follows:

The FWI method for zero-offset borehole data involves estimating the initial model based on borehole rock core samples and the source wavelet through controlled experimental water tank tests. A low-pass filter with an increasing high-cut frequency is applied to preprocess the data before each iteration. The inversion is performed in a frequency-dependent manner, starting from the low-frequency range and progressing to the high-frequency range. The envelope-based FWI mitigates local minima and provides a superior initial model for the correlation-based FWI. The resulting inverted S-wave velocity profile enables the identification of karst voids outside the borehole.

Synthetic examples of three typical karst void scenarios (single void, multiple voids, and linear fracture) are used to validate the proposed FWI method. The envelope-based FWI effectively incorporates low-frequency information into the spectrum, addressing the issue of local minima in zero-offset data. The correlation-based FWI significantly enhances the inversion results based on the model provided by the envelope-based FWI. The application of a threshold value on the S-wave velocity aids in the identification of karst void anomalies.

A field experiment is conducted on a bridge pile foundation project in a karst area to assess the applicability of the proposed FWI method. Single borehole data is used to detect karst voids at the pile location and provide geological information for pile bearing layer design. The inverted S-wave velocity profile is compared with cross-hole P-wave velocity ultrasonic tomography, revealing the effectiveness of the proposed FWI method in accurately reconstructing karst voids' shapes, depths, and infill. However, the FWI lacks azimuthal information.

The FWI of zero-offset borehole data has practical significance in engineering applications, particularly in pile, tunnel, and mining projects. Borehole investigations are crucial for understanding potential anomalies around a borehole, influencing the design and construction processes of structures. The intuitive imaging features of the FWI method make it highly applicable in a wide range of engineering problems.

## References

Bohlen T (2002) Parallel 3-D viscoelastic finite difference seismic modelling. Comput Geosci 28(8):887–899

Bozdag E, Trampert J, Tromp J (2011) Misfit functions for full waveform inversion based on instantaneous phase and envelope measurements. Geophys J Int 185(2):845–870

Bunks C, Saleck FM, Zaleski S et al (1995) Multiscale seismic waveform inversion. Geophysics 60(5):1457–1473

Chi B, Dong L, Liu Y (2014) Full waveform inversion method using envelope objective function without low frequency data. J Appl Geophys 109:36–46

Choi Y, Alkhalifah T (2012) Application of multi-source waveform inversion to marine streamer data using the global correlation norm. Geophys Prospect 60(4):748–758

Groos L, Schaefer M, Forbriger T et al (2017) Application of a complete workflow for 2D elastic full-waveform inversion to recorded shallow-seismic Rayleigh waves. Geophys J Society of Explorat Geophys 82(2):R109–R117

Köhn D, Meier T, Fehr M et al (2016) Application of 2D elastic Rayleigh waveform inversion to ultrasonic laboratory and field data. Near Surf Geophys 14(5):461–476

Komatitsch D, Martin R (2007) An unsplit convolutional perfectly matched layer improved at grazing incidence for the seismic wave equation. Geophysics 72(5):SM155–67

Kurzmann A (2012) Applications of 2D and 3D full waveform tomography in acoustic and viscoacoustic complex media. PhD thesis. Karlsruhe Institute of Technology (KIT)

Liu H, Maghoul P, Shalaby A, Bahari A, Moradi F (2020) Integrated approach for the masw dispersion analysis using the spectral element technique and trust region reflective method. Comput Geotech 125:103689

Pan Y, Gao L, Bohlen T (2019) High-resolution characterization of near-surface structures by surface-wave inversions: from dispersion curve to full waveform. Surv Geophys 40(2):167–195

Pan Y, Gao L, Shigapov R (2020) Multi-objective waveform inversion of shallow seismic wavefields. Geophys J Int 220(3):1619–1631

Robertsson JO, Blanch JO, Symes WW (1994) Viscoelastic finite-difference modeling. Geophysics 59(9):1444–1456

Smith JA, Borisov D, Cudney H et al (2019) Tunnel detection at Yuma Proving Ground, Arizona, USA—Part 2: 3D full-waveform inversion experiments Tunnel detection at YPG—Part 2: 3D FWI. Geophysics 84(1):B95–B108

Taner MT, Koehler F, Sheriff RE (1979) Complex seismic trace analysis. Geophysics 44(6):1041–1063

Tang X, Zheng Y, Patterson D (2007) Processing array acoustic-logging data to image near-borehole geologic structures. Geophysics 72(2):E87–E97

Tarantola A (1984) Inversion of seismic reflection data in the acoustic approximation. Geophysics 49(8):1259–1266

Tran K, McVay M, Faraone M et al. (2013) Sinkhole detection using 2D full seismic waveform Tomography. Geophysics 78(5):R175–R183

Virieux J (1986) P-SV wave propagation in heterogeneous media: velocity-stress finite difference method. Geophysics 51(4):889–901

Wang Y, Miller RD, Peterie SL et al (2019) Tunnel detection at Yuma Proving Ground, Arizona, USA—Part 1: 2D full-waveform inversion experiment Tunnel detection at YPG—Part 1: 2D FWI. Geophysics 84(1):B95–B105

Wu R, Luo J, Wu B (2014) Seismic envelope inversion and modulation signal model. Geophysics 79(3):WA13–24

# Chapter 6
# Borehole Radar Detection Method

## 6.1 Prologue

Borehole radar (Zhong et al. 2011 and 2012; Ma et al. 2014 and 2015; Ebihara 2004) is an emerging method in geophysics used for subsurface investigations. It can be employed in two ways: cross-hole tomography and single borehole reflection survey. The former allows for the estimation of velocity models and attenuation coefficients between two boreholes. Conductivity and permittivity of the stratum can be determined using techniques like full-waveform inversion and Bayesian inversion (Qin et al. 2018; Liu et al. 2019). On the other hand, single borehole radar is effective in identifying geological anomalies away from the borehole walls (Liu et al. 2017) It offers advantages over cross-hole detection in terms of time and cost, making it a feasible option for studying karst anomalies at specific locations. Time–frequency analysis and filtering techniques (Liu et al. 2017; Alperovich et al. 2013) are employed to process GPR data and enhance weak signals amidst noise.

This chapter focuses on the development of a specialized prototype system for detecting karst voids at pile locations using a near offset and small diameter single borehole GPR. This system serves as a supplementary method to borehole sonic detection. To enhance the detection capabilities, a rapid data processing approach combining Hilbert transform (HT) for complex signal analysis and medium filtering was proposed. This method effectively extracts weak backscattered waves from distant geological reflectors and facilitates the identification of karst void infill. The applicability of the developed prototype and the proposed data analysis method were demonstrated through a controlled horizontal borehole experiment. Furthermore, field single borehole radar detections were carried out in an urban karst area during a freeway bridge extension project in Guangzhou, China. The obtained data from various karst geological anomalies were analyzed, allowing for insights into the construction methods of piles and the remediation techniques for different karst anomalies. These conclusions were based on the detection results and feedback received during ongoing construction activities.



L. Liu et al., *Detection of Karst Voids at Deep Pile Foundation*,
https://doi.org/10.1007/978-981-97-5834-0_6

## 6.2 Single Borehole Radar Detection

### 6.2.1 Detection Scheme

Karst features such as sinkholes, boulders, and karst voids occur randomly at pile locations, as depicted in Fig. 6.1a, and have a significant impact on both the bearing capacity and construction safety. The bearing capacity of a pile relies on the contributions from skin friction and end bearing. In the presence of sinkholes or karst voids within the depth range (e.g., the sinkhole and three shallow karst voids in Fig. 6.1a, the slide skin friction should be entirely disregarded when calculating the pile's bearing capacity. In karst areas, the end bearing capacity assumes greater importance. However, the capacity of the pile is considerably reduced due to various karst anomalies, such as voids, fractures, and regional rock dissolution beneath the pile (e.g., the deepest karst void in Fig. 6.1a). In terms of construction safety, unknown karst anomalies at the pile location can give rise to four potential incidents.

Stuck Drill: When drilling penetrates the thin roof of karst voids, particularly with long and heavy stroke distances, it poses a grave danger.

Slurry Leakage: The drilling process can lead to the drainage of slurry from the pile hole if it penetrates karst voids or fracture flow tunnels.

Inclined Pile Hole: The drilling of uneven or irregular geological interfaces increases the risk of inclined pile holes.

Pile Hole Collapse: Neglecting appropriate measures to address the aforementioned incidents can result in the collapse of a pile hole.

Properly addressing these challenges is of utmost importance to ensure construction safety and minimize the potential risks posed by karst anomalies at pile locations. Neglecting to detect and address karst voids can be extremely hazardous as it may result in inadequate bearing capacity design and construction practices. To mitigate these risks, the utilization of single borehole GPR detection (as shown in Fig. 6.1b) becomes crucial. This detection method involves probing the entire diameter of the pile in conjunction with the existing borehole, allowing for more comprehensive geological information that can be utilized in designing appropriate bearing capacity and construction methods.

### 6.2.2 Single Borehole GPR Prototype System

The pulse GPR system, depicted in Fig. 6.2, is designed as a single borehole prototype. It comprises several components, including a surface computer, transmitting and receiving antenna, fiber cable, depth counter, and charger. The downhole antenna is connected to the computer via a fiber cable, which passes through the depth counter to record the probe's depth. The transmitting antenna is constructed as an omnidirectional dipole, consisting of two conical copper tubes, and operates within a frequency range of 20 to 200 MHz. Similarly, the receiving antenna mirrors the structure of the transmitting antenna and incorporates a 16-bit A/D converter. During testing, the borehole GPR downhole transmitting and receiving antenna, as illustrated in Fig. 6.2, is gradually lowered from the top to the bottom of the borehole at a velocity of 0.1 m/s. This process is facilitated using the ground support depicted in Fig. 6.1b. Electromagnetic (EM) waves are transmitted at intervals of 0.1 m in depth. These waves propagate through the borehole and along its path, reflecting off impedance interfaces. Subsequently, the waves are received and preprocessed by the receiving antenna. The in situ computer then presents and analyzes the EM wave profile.

The system incorporates two crucial features for detecting karst voids at pile locations: a small diameter (36 mm) and near offset (30 cm). These factors play a

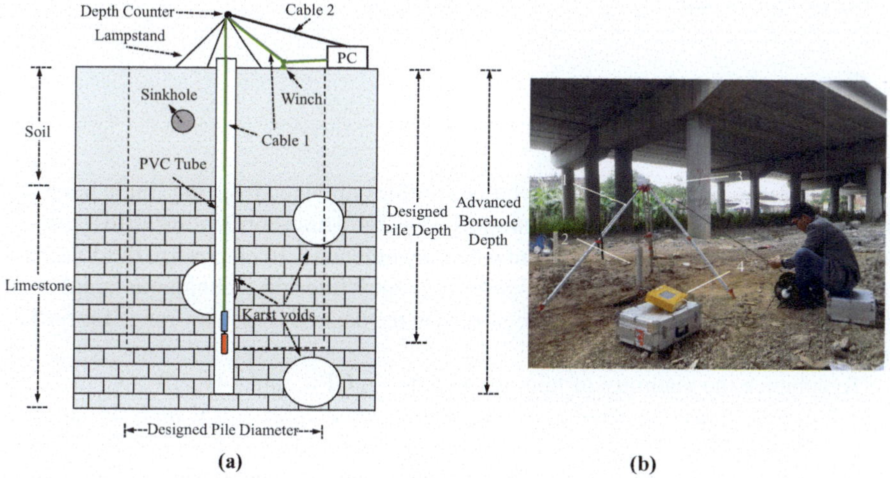


**Fig. 6.1** **a** Schemaic diagram of the single borehole detection deployed in a borehole at pile location. Multiple karst voids within the pile are depicted by a dashed line. **b** Single borehole test in the field. PVC casing is placed in the borehole to avoid borehole collapse and shrinkage. 1—lampstand, 2—borehole with PVC casing, 3—Depth counter, 4—PC

**Fig. 6.2** The single borehole ground penetration radar (GPR) prototype system. 1—computer, 2—transmitting antenna, 3—receiving antenna, 4—fiber cable, 5—depth counter, 6—charger

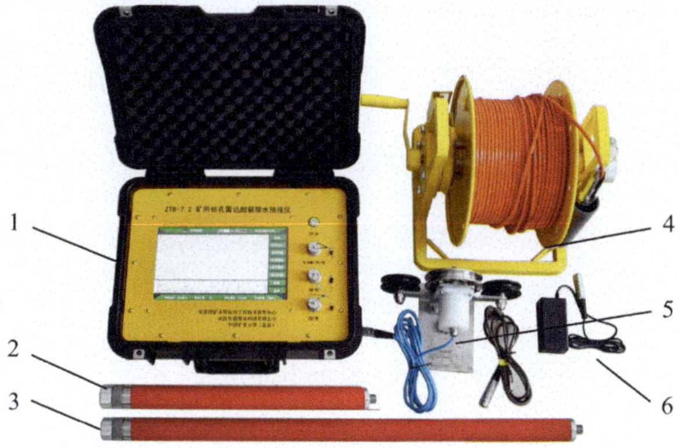

significant role in the effectiveness of the detection process. In karst areas, where the drilling borehole diameter is typically around 70 mm, it is necessary to utilize PVC tubes during borehole GPR detection. These tubes help prevent the infill of voids, as well as unstable and disturbed rocks, from falling into the borehole and potentially interfering with the antennae. PVC tubes are installed after drilling, making the device easier to operate, particularly when the diameter of the tubes is small. This small diameter downhole tool proves more practical for civil engineering projects. Furthermore, in karst areas, geological anomalies can vary significantly even with slight changes in depth. The near offset acquisition system employed in the detection process ensures high resolution. It enables the use of a medium filter to effectively remove direct waves and borehole reflection background noise, enhancing the clarity of the obtained data.

### 6.2.3 Signal Analysis

During the signal analysis of the EM wave profiles obtained from the single borehole system, the static correction can be simplified as zero-time correction due to the small interval between the transmitting and receiving antennas. To emphasize the dominant frequency of the antennas, a band-pass frequency filtering technique is employed. This filtering approach helps to concentrate on the desired frequency range for further analysis.

To enhance the visibility of faint backscatter in GPR data, two effective methods are employed: gain control based on amplitude and complex signal analysis based on instantaneous phase. The use of instantaneous phase analysis proves more suitable for processing GPR data obtained from borehole tests in karst areas. This is because the saturated soil and rocks found in karst regions tend to significantly attenuate the energy of the EM wave, making the amplitude-based analysis less reliable. The Hilbert transform is used to calculate the instantaneous phase. Taking one depth test signal $s_x(t)$ as an example, an analytic signal $z_x(t)$ is constructed, which includes the real component $s_x$(t) and the imaginary component $y_x(t)$ that can be constructed using the HT (Taner et al. 1979):

$$z_x(t) = s_x(t) + iy_x(t), \tag{6.1}$$

where the imaginary component $y_x(t)$ is obtained via the Hilbert transform of $s_x(t)$:

$$y_x(t) = H(s_x(t)) = \frac{1}{\pi} \int_{-\infty}^{+\infty} \frac{s_x(\tau)}{t - \tau} d\tau, \tag{6.2}$$

where $H(s_x(t))$ is the Hilbert transform of $s_x(t)$. The instantaneous phase $\phi_x(t)$ of the signal $s_x(t)$ can be obtained as follows:

$$\phi_x(t) = \arctan\frac{y_x(t)}{s_x(t)} \tag{6.3}$$

To extract the effective reflected waves and remove borehole background noise, median filtering can be used. The single borehole EM wave profiles, $P(t, x)$, typically contain direct waves, waves in the borehole, and reflected waves from the geological structure outside the borehole. The direct waves and the multiple reflected EM waves in the borehole exhibit constant and small arrival times because of the near offset acquisition system. However, most backscatter EM waves from the geological structure have sloping and variable time appearance since the wave travel distance between the probe of the geological structures varies along the borehole. Thus, median filtering can be used to remove the constant time arrivals while preserving variable time. The procedure of the median filter is as follows:

(a) Taking the test signal at one depth $s_x(t)$ as an example, a new set of sequences can be formed by reading the test signals of 2N + 1 channels at adjacent depths ($s_{x\text{-}N}(t)$, $s_{x\text{-}N+1}(t)$... $s_{x+N}(t)$).
(b) For every point at time $s_x(t_n)$, the amplitudes of the test signals of 2N + 1 channels ($s_{x\text{-}N}(t_n)$, $s_{x\text{-}N+1}(t_n)$... $s_{x+N}(t_n)$) are reordered. The median $s_{xm}(t_n)$ replaces the original amplitude $s_x(t_n)$ in the test signal $s_x(t)$.
(c) Repeat steps a) and b) for all the recordings. A new wave can be obtained as a $P_m(t, x)$ profile after median filtering, which consists of the direct wave and waves in the borehole.
(d) The wave profiles $P_r(t, x)$ can be calculated as $P_r(t, x) = P(t, x) - P_m(t, x)$, which mainly contain the reflected waves from the geological structure

## 6.3 Validation by a Controlled Experiment

The single borehole radar system and its signal analysis method underwent testing in a field experimental setup involving horizontal boreholes. A series of controlled tests were conducted, which included the following aspects: (1) Analysis of the typical wave field and profile. (2) Identification of karst points and linear fracture anomalies. (3) Assessment of the performance of instantaneous phase and medium filtering.

### *6.3.1 Experiment Setting*

A controlled horizontal borehole experiment was carried out in a coal mine located in Datong Mine, Chongqing, China, specifically at a roadway crossing. The surrounding geological formation in this area consists of saturated limestone, which aligns with the lithology commonly found in karst regions. The experimental setup is depicted in Fig. 6.3. In this experiment, a horizontal borehole was drilled within roadway 1 to generate an image of roadway 2 and a shelter room located in roadway 2 using single

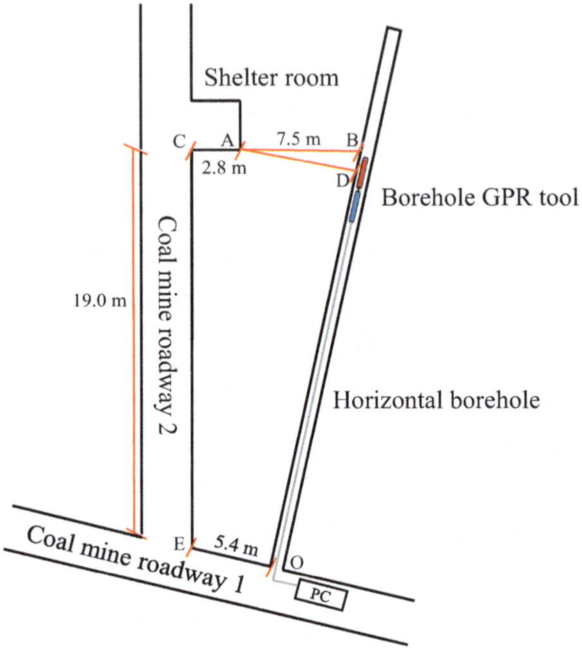


**Fig. 6.3** Geometry setup of controlled single borehole GPR test. AB, AC, CE, and EO separation is 7.5, 2.8, 19.0, and 5.4 m, respectively, according to the roadway design. AD, BD, and DO separation can be calculated as 7.28, 1.81, and 19.12 m, respectively

borehole GPR detection. The aim was to simulate three typical geological structures of interest to engineers in karst areas: a soil-rock interface, a linear rock fracture, and a point-shaped karst void. These structures were represented by the interfaces between roadway 1, roadway 2, and the shelter room, respectively. Due to the attenuation of high-frequency electromagnetic waves in saturated rock, a pair of borehole antennae with a relatively lower frequency of 30 MHz was utilized in the controlled test. This frequency choice was made considering security concerns, as large energy transmission is restricted within the mine environment. The controlled horizontal borehole experiment provided valuable insights into the application of single borehole GPR detection in karst areas, particularly in understanding the imaging capabilities and limitations in detecting and characterizing different geological structures.

### *6.3.2 Results*

The EM wave profiles obtained from the controlled horizontal borehole experiment are illustrated in Fig. 6.4. In the original data (Fig. 6.4a), multiple constant arrival events can be observed. These events correspond to direct wave events and multiple reflections within the borehole. They present as continuous background noises throughout the depth of the borehole. A subtle discontinuity is noticeable at a depth of 8.5 m, which corresponds to a change in saturation. In the shallower

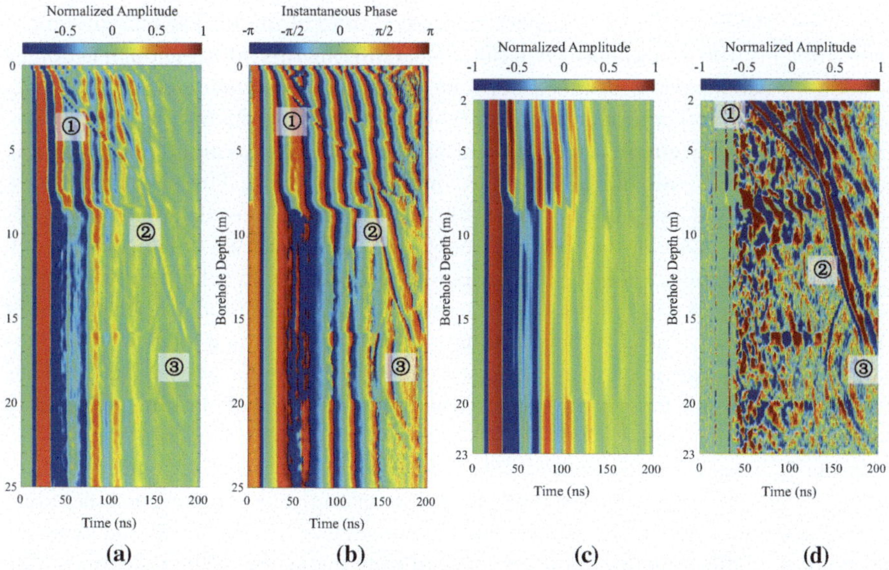


**Fig. 6.4** Original single borehole GPR detection data of controlled horizontal borehole experiment: **a** processed radar data; **b** instantaneous phase; **c** removed wave profile using median filtering; **d** median filtered wave. Two meters at the top and bottom are removed during median filtering

section, where the rock formation is drier, the velocity of EM waves is slightly higher compared to the deeper and more saturated formation. Within Fig. 6.4a, two weak sloping events are visible, albeit barely discernible. Event 1 is relatively clear, whereas Event 2 is less distinct as the amplitude of the reflected wave is faint. If the gain control is not appropriately set, Event 2 may be overlooked or deemed insignificant. Event 3 represents hyperbolic reflected waves, although the backscattered energy associated with it is weak.

By analyzing the instantaneous phase of the EM wave profiles presented in Fig. 6.4b, distant events become more prominent. Event 1 remains clearly visible at the end of the sampling time (200 ns and 10.5 m depth). Event 1 signifies guided waves that propagate along the borehole and undergo reflection at the interface characterized by a change in permittivity. This interface corresponds to the surface of the coal mine roadway 1 depicted in Fig. 6.3. Consequently, it becomes possible to estimate the velocity of the EM waves in the formation using the following approach:

$$v = 2d/t = 2 \times 10.5\,m/200\,ns = 0.105\,m/ns, \tag{6.4}$$

The estimated velocity aligns with the known EM wave velocity in limestone, which provides consistency and reliability in the analysis. Considering this velocity, Event 2 can be interpreted as the backscattered waves originating from the surface of coal mine roadway 2. Similarly, Event 3 corresponds to the backscatter originating from the shelter room located on roadway 2. An interesting observation is that at the entrance of the borehole (0 m depth), the arrival time of the backscattered wave is approximately 100 ns, as depicted in Fig. 6.4b. Utilizing this information, it becomes possible to calculate the distance between the borehole entrance and roadway 2, referred to as EO separation in Fig. 6.3, which amounts to 5.25 m.

To identify the accurate arrival time of backscattered waves and reduce borehole background noise, a medium filtering technique is employed. This filtering process effectively removes unwanted waves, as illustrated by the removed and filtered waves displayed in Fig. 6.4b. In the case of Event 3, the earliest arrival time is observed at 137 ns (18.1 m depth). Using this information, it is possible to calculate the distance between the nearest point in the borehole (point D in Fig. 6.3) and the nearby corner of the shelter room. The calculated distance amounts to 7.19 m, which closely aligns with the actual position of point D. It is important to note that slight discrepancies may arise due to variations in the EM wave velocity near the shelter room, particularly in the concrete structure.

Event 2, observed at a depth of 18 m and depicted in Fig. 6.4d, corresponds to the backscatter originating from roadway 2. Interestingly, all three events, including Event 2, exhibit a lack of phase reversal. This characteristic aligns with the expected behavior of air-filled anomalies, further confirming the nature of the reflections. The controlled experiment conducted demonstrates that a single borehole GPR detection can yield three typical backscattered events. In Fig. 6.4, Event 1 represents a linear backscattered event associated with a geological interface crossing the borehole, such as the contact between regolith and bedrock or a rock fracture. The slope of this event can be utilized to estimate the velocity of the EM wave. Event 2 corresponds to linear rock fractures situated outside the borehole, and the slope of this event can provide insights into the shape and trend of the fracture. Event 3, on the other hand, exemplifies a typical hyperbolic backscatter, indicating the presence of a point-shaped geological anomaly like karst voids, sinkholes, or boulders. By employing instantaneous phase analysis and medium filtering, the faint events are accentuated, making the phase of the backscatter more visible. This enhancement proves beneficial in determining the composition of the anomaly's infill.

## 6.4 Field Test

In Guangzhou, a major city in southern China, single borehole GPR detection tests were conducted as part of a freeway extension project. The objective of the project is to replace a four-lane road with a new eight-lane freeway, as illustrated in Fig. 6.5a. The overall length of the project spans 57.6 km, with approximately 33 km situated in a karst area. This karst region encompasses over 3000 piles, and the presence of underground karst voids poses risks to the bearing capacity of the piles, as well as the safety of construction activities and nearby structures. The study area primarily consists of Quaternary regolith, with a thickness ranging from approximately 15 to 25 m. The overburden layer consists of plain fill, silty clay, and silty sand. The underlying bedrock layer is composed of medium-weathered and weakly weathered

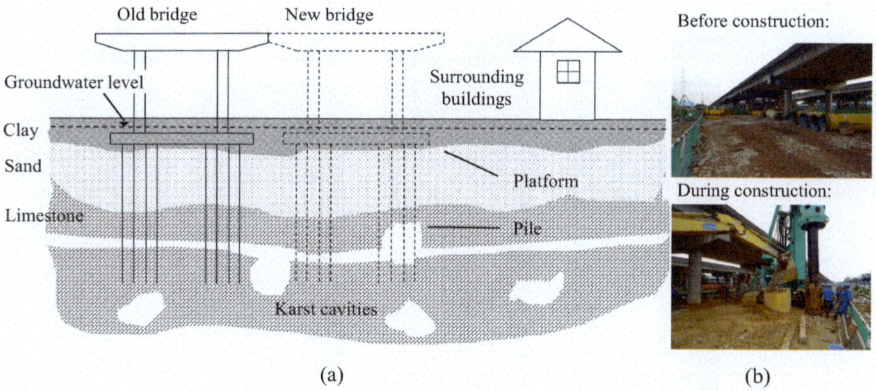


**Fig. 6.5** Schematic of the freeway extension project. **a** Simplified stratigraphy of the freeway extension bridge project. A new bridge will be built immediately adjacent to the old bridge; **b** Field site before and during construction

limestone. Karst cavities are prevalent in the project area, adding to the complexity of the geological conditions. To identify and assess the karst geological anomalies at the pile locations, single borehole GPR detection was employed. This approach enabled the investigation of various scenarios encountered at different pile locations. Specifically, four typical scenarios are discussed in this section: karst voids (ZK-4-11 borehole, located at the bridge project Sect. 6.4), sinkholes (ZK-2-23 borehole), rock fractures (ZK-7-13 borehole), and integrated rock (ZK-4-36 borehole). These examples provide insights into the diverse geological conditions encountered and the potential challenges faced during the project.

### 6.4.1 Test 1: Karst Voids

At a planned pile location, a borehole identified as ZK-4–11 was drilled, revealing a significant karst void measuring 1 m in size between depths of 29 and 30 m. Additionally, a soil-rock interface was observed at a depth of 24 m. The stratigraphy of the borehole (depicted in Fig. 6.6a indicates that the limestone is weakly weathered, with over 60 RQD values. According to the borehole data, the pile should be designed to be positioned directly beneath the floor of the karst void, at approximately 30 m in depth. However, single borehole GPR detection at the pile location has identified multiple karst geological anomalies.

The radar data obtained from the GPR detection were processed through zero-correction and band-pass filtering, resulting in the representation shown in Fig. 6.6b. Event 1 demonstrates the discontinuation of direct EM waves and borehole reflections at a depth of 29 m, consistent with the borehole stratigraphy. Events 2 and 3 display linear features intersecting at a borehole depth of 28 m, suggesting the presence of

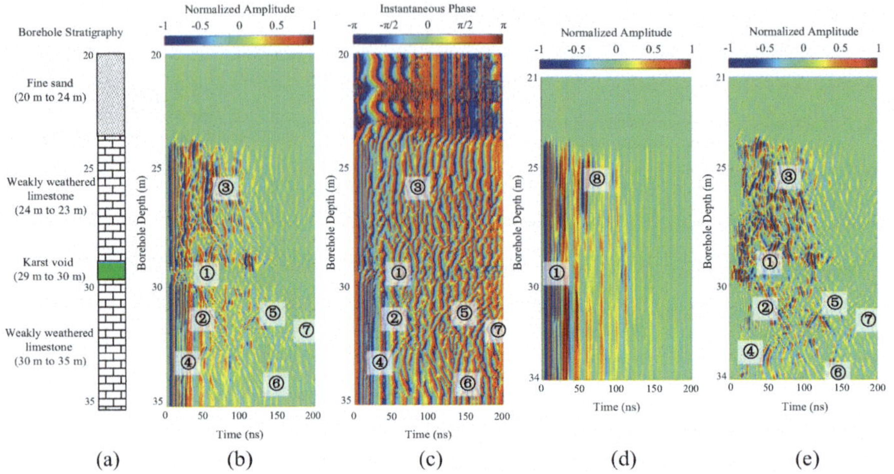


**Fig. 6.6** Single borehole GPR detection data for a typical karst void borehole scenario. **a** No. ZK-4–11 borehole stratigraphy; **b** radar data; **c** instantaneous phase; **d** removed wave profile using median filtering; **e** median filtered wave

a rock fracture that was missed in the borehole stratigraphy. Event 4 corresponds to hyperbolic reflected waves, indicating the presence of a karst cavity surrounding the borehole. Similar hyperbolic events, observed in Events 5, 6, and 7, appear at depths ranging from 30 to 35 m, indicating the existence of multiple karst voids at the pile location.

Figure 6.6c-e present the instantaneous phase and median filtered data, offering more detailed information. In Fig. 6.6b, the backscatter behind Event 1 is not visible due to the significant attenuation of EM waves in the water or saturated soil-filled karst cavity. However, in the instantaneous phase (Fig. 6.6c) and the removed waves profile (Fig. 6.6d), the borehole reflection behind Event 1 can be identified. Event 2 and Event 3 are more pronounced in the instantaneous phase and median filtered waves (Fig. 6.6e). Event 2 discontinues at a depth of 30 m and exhibits a consistent linear trend from 28 to 35 m. The slowly decaying backscattered wave from the rock fracture at 28 m suggests that the surrounding rock is integrated in a certain direction. The slope of the event allows estimation of the EM velocity in the limestone, which is calculated to be 0.112 m/ns. Consequently, Events 4, 5, 6, and 7 indicate the presence of three karst cavities located approximately 2.2 m, 5.7 m, and 7.5 m from the borehole. Events 6 and 7 are continuous in the instantaneous phase, indicating that they correspond to the same cavity. Phase reversal in Events 4 to 7 can be observed in the median filtered wave (Fig. 6.6e), indicating that these three cavities are filled with water or saturated soil. Moreover, Fig. 6.6d reveals an amplitude anomaly at a depth of 25 m (Event 8), which corresponds to a karst void with an outline parallel to the borehole. Based on the results of the single borehole GPR detection, it is deemed unsafe to design the bearing layer at a depth of 30 m. Instead, the borehole should be deepened to explore another integrated bedrock layer.

Regarding the feedback received from the pile construction of borehole ZK-4–11 (scenario 1), a steel casing was employed to protect the pile hole from collapse, reaching a depth of 35 m. The foot blade of the casing was reinforced at depths of 24 m (soil-rock interface), 28 m (a rock fracture identified by borehole radar detection), 29 m and 30 m (the top and floor, respectively, of the karst void identified by the borehole), and 33 m and 35 m (the top and floor, respectively, of the karst void detected by borehole radar). Additionally, artificial construction materials were deposited, mixed, and compacted at depths of 24 m, 29 m, and 33 m to address the karst voids encountered.

### 6.4.2 Test 2: Sinkhole

The stratigraphy of borehole ZK-2-23 reveals the presence of a karst cavity measuring 1 m in diameter at a depth of 31 to 33 m. The overlying layers consist of artificial filling, fine sand, and silty clay. The upper bedrock is medium-weathered limestone with an approximate RQD value of 30, while the lower layer is weakly weathered limestone with an RQD value exceeding 60. The original GPR data obtained from the borehole align with the observations made from the bedrock stratigraphy.

The GPR data further detect a weak direct wave indicative of the karst void in the depth range of 31 to 33 m (referred to as Event 1 in Fig. 6.7). Notably, the direct wave events between 25 and 31 m exhibit greater irregularity compared to those between 33 and 39 m, suggesting increased weathering and dissolution of the rock within the 25–31 m range. By utilizing the Hilbert transform and medium filtering, reflected EM waves from the geological structure behind the borehole are identified as Events 3, 4, 5, and 6. Event 3 corresponds to a linear rock fracture intersecting the borehole at a depth of 30 m. Events 4 and 5 represent hyperbolic backscattered waves originating from two separate karst voids. The estimated distances between the borehole and these voids are approximately 4.4 and 7.0 m, respectively.

In contrast to the data obtained from scenario 1, the waveform observed in scenario 2 does not exhibit a frequency decrease within the soil layer. Instead, weak high-frequency signals (referred to as Event 2, approximately 300 MHz) can be observed in the original borehole GPR data. Repeated tests have confirmed that these high-frequency signals are not random background noise but rather ringing or reverberation (300 MHz) from signals present within the borehole.

To extract meaningful events, FIR low-pass filtering is applied. Notably, in the low-pass filtered data shown in Fig. 6.7f, a faint hyperbolic reflected wave event can be observed at a depth of 12 m. The slope of the hyperbola becomes more pronounced after conducting instantaneous phase analysis and median filtering, as depicted in Fig. 6.7 g and h. Based on the slope of the hyperbola, the estimated EM velocity is determined to be 0.034 m/ns, indicating that the EM waves are propagating in water and that the reflector is approximately 0.6 m away from the borehole. However, it is important to note that the borehole diameter is 0.1 m. The enlargement of the borehole is attributed to drilling disturbances and the loss of surrounding soil. Given

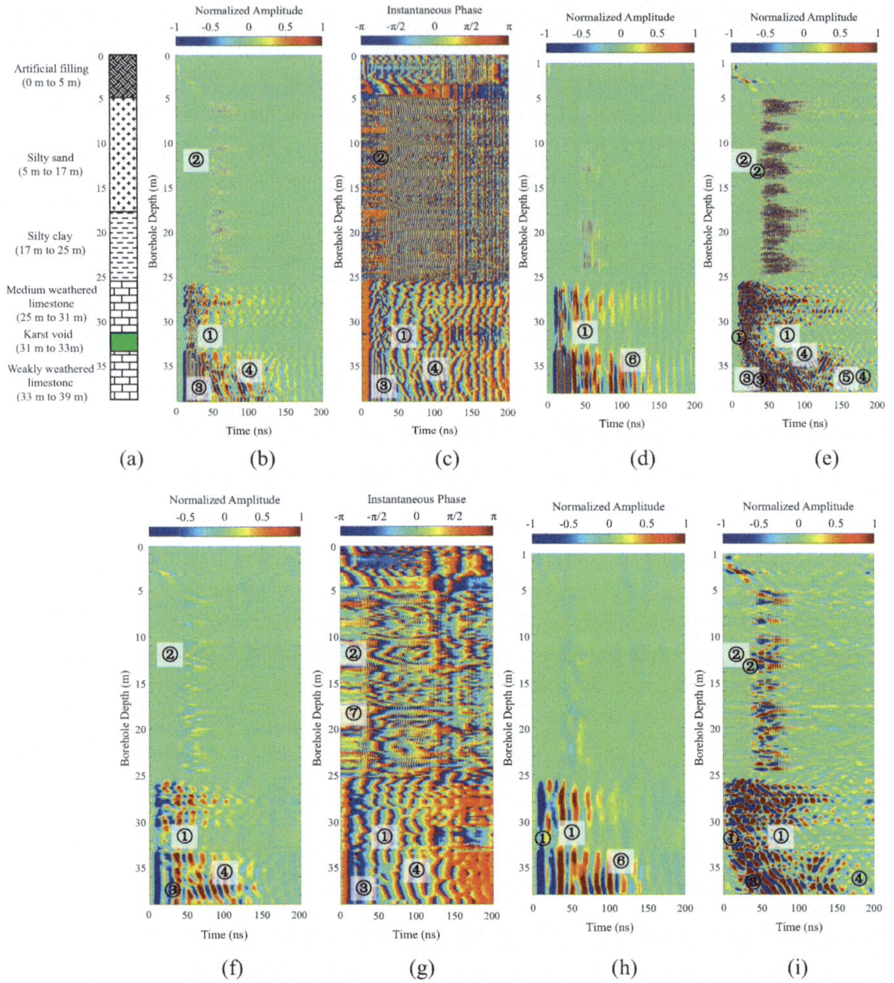


**Fig. 6.7** Single borehole GPR detection data on a typical karst void borehole scenario. **a** No. ZK-2-23 borehole stratigraphy; **b** original radar data; **c** instantaneous phase; **d** removed wave profile using median filtering; **e** median filtered wave; **f** Low-pass filtered radar data; **g** instantaneous phase of low-pass filtered radar data; **h** median filtering removed wave profile of low-pass filtered data. **i** median filtered wave of low-pass filtered data

the presence of a karst void in the bedrock, the loss of covering materials in the silty sand and clay layer, and the presence of a medium-weathered limestone layer providing a pathway for soil to funnel into the karst void, this borehole poses a high risk of sinkhole formation. Therefore, sinkhole remediation measures are necessary before proceeding with construction.

To address the sinkhole, compaction grouting was carried out within a depth range of 10 to 25 m, covering the soil-rock interface up to the top of the sinkhole, prior to the start of construction. The sinkhole was treated with a mixture of cement, sodium silicate, crushed stone, and red clay, which was deposited, mixed, and compacted at depths of 25 m and 31 m. During the pile construction process, a steel casing was lowered to a depth of 37 m and reinforced at critical interfaces located at depths of 25 m, 31 m, and 33 m. These measures were implemented to mitigate the risks associated with the sinkhole and ensure the stability of the construction project.

### 6.4.3 *Test 3: Rock Fracture*

From the borehole stratigraphy of ZK-7-13 (depicted in Fig. 6.8a), it can be determined that weakly weathered limestone is present within the depth range of 20 to 42 m. The RQD values obtained from the drilling core indicate values exceeding 70. Although the borehole is situated in an area where karst development is rare, minor limestone dissolution can still be observed in this specific field area. Single borehole radar detection was conducted to confirm the absence of any hazardous geological structures near the planned pile location. The recorded GPR signals are displayed in Fig. 6.8a, where the nearly coincident arrival times and amplitudes of the direct waves suggest the presence of an integrated rock formation behind the borehole.

However, the GPR wave profile also exhibits a notable reflection within the depth range of 27 to 36 m. The backscattered waves illustrate a hyperbolic event at the beginning (referred to as Event 1 in Fig. 6.8a), which then transitions into a linear Event 2. The shape of this event suggests the existence of a point-shaped karst void anomaly at a depth of 27 m, connected to a rock fracture within the depth range

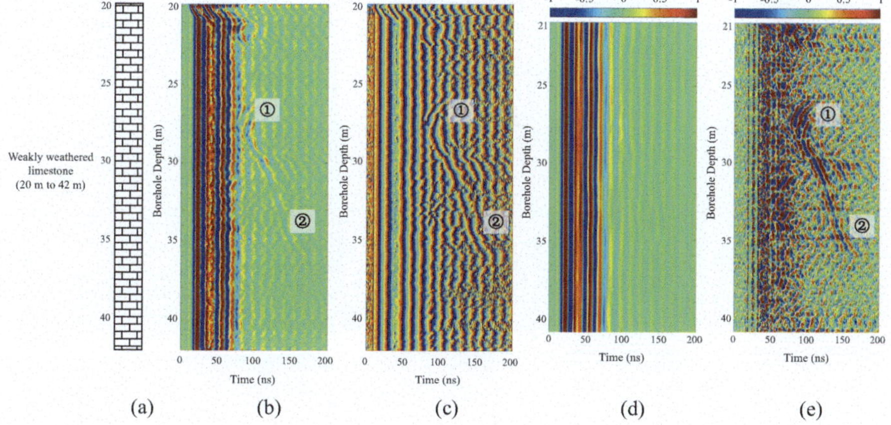


**Fig. 6.8** Single borehole GPR detection data on a typical rock fracture borehole scenario. **a** No. ZK-7–13 borehole stratigraphy; **b** radar data; **c** instantaneous phase; **d** removed wave profile using median filtering; **e** median filtered wave

of 30 to 36 m. Applying the Hilbert transform and medium filtering allows for the separation of the borehole background and effective backscatter. Phase reversal of the backscatter can be observed, indicating that both the karst void and rock fracture are filled with water or saturated soil. The estimated distance between the karst void and the borehole amounts to approximately 4.5 m. The bedrock range of 35 to 40 m is considered suitable for establishing a bearing layer for the pile. During the construction, the steel casing was only lowered to the soil-rock interface at a depth of 20 m.

### *6.4.4 Scenario 4: Integrated Bedrock*

The borehole analysis of ZK-4-36 reveals the presence of a 3 m thick layer of artificial filling, followed by a 9 m thick layer of medium sand, 4 m of silty clay, 7 m of silty fine sand, and weakly weathered limestone exceeding 10 m in thickness. The borehole GPR data, as depicted in Fig. 6.9b, demonstrate the rapid attenuation of EM wave amplitudes within the soil layer. Direct waves can be observed in the depth range of the sand layer but are not visible within the clay layer. The direct waves within the bedrock range exhibit earlier arrival times compared to those within the soil layer, owing to the faster EM wave velocity in limestone. The waveform beyond the borehole generally appears regular, although visual identification of arrival times and event shapes for reflections in the upper bedrock (Event 1 and 2 in Fig. 6.9b) proves challenging.

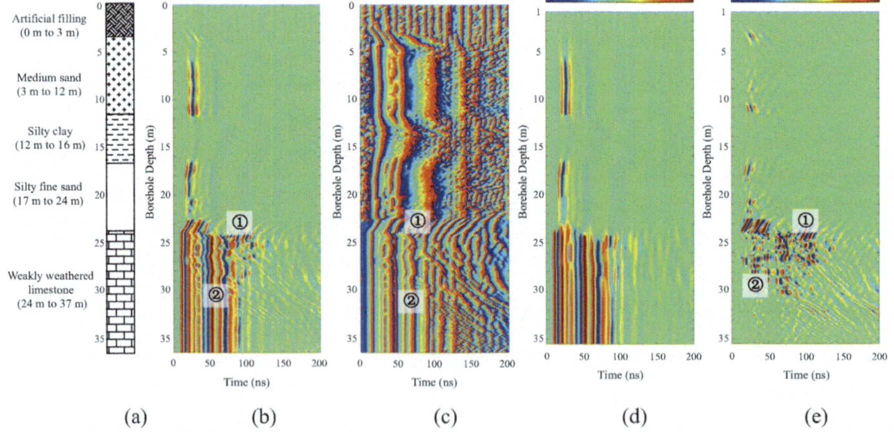


**Fig. 6.9** Single borehole GPR detection data on a typical rock fracture borehole scenario. **a** No. ZK-4-36 borehole stratigraphy; **b** radar data; **c** instantaneous phase; **d** removed wave profile using median filtering; **e** median filtered wave

In Fig. 6.9c, the presence of direct waves and subsequent waveforms within the silty clay layer, which has a slower velocity than the sand layer, can be observed. The instantaneous phase analysis indicates a lower signal frequency within the soil layer due to the rapid attenuation of high-frequency signals. Median filtered waves (Fig. 6.9e) suggest that Event 1 corresponds to the soil-rock interface at a depth of 24 m. Based on the slope of this event, the estimated EM wave velocity is 0.114 m/ns, consistent with the velocity of the weakly weathered limestone. Event 2 indicates the presence of a linear rock fracture at a depth of 26 m. No visible backscatter is observed within the depth range of 26 to 36 m, suggesting an integrated nature of the bedrock in this range. During the pile construction, the steel casing was lowered to the soil-rock interface at a depth of 24 m.

## 6.5 Summary

In this chapter, we utilize a single borehole radar detection system to investigate geological anomalies at pile locations. The detection method was tested in a controlled horizontal borehole setup and subsequently applied in a freeway bridge extension project situated in an urban karst area. The following key findings can be drawn.

Karst geological anomalies exert a significant impact on the bearing capacity and construction of piles. To effectively detect karst voids at pile locations, a specially designed single borehole GPR prototype with a small diameter (36 mm) and near offset (30 cm) was developed. The prototype allows for high-resolution imaging of karst anomalies. The application of instantaneous phase analysis through Hilbert transform proves effective in enhancing low-energy reflections, while medium filtering helps distinguish weak reflections from borehole background noise.

A controlled single borehole GPR experiment was conducted in a mining horizontal borehole to validate the prototype and assess the advantages of signal analysis methods. The experiment yielded three typical reflection events in the detection of karst anomalies at pile locations: (a) a linear event intersecting the borehole corresponding to a soil-rock interface or rock fracture crossing the borehole; (b) a linear event outside the borehole indicating a linear rock fracture external to the borehole; and (c) a hyperbolic reflection signifying point-shaped karst voids or sinkholes.

Single borehole GPR detection was employed in the freeway bridge extension project located in an urban karst area. The results highlight the rapid decay of high-frequency signals in the soil and in the infill depth range of karst voids. The direct EM wave event exhibits discontinuity and irregularity within the depth range of weathered rock. The velocity of EM waves can be determined by analyzing the slope of linear fractures that intersect the borehole. The location of surrounding karst voids can be estimated by assessing the velocity and arrival time of hyperbolic reflections. Phase reversal may indicate the presence of water or saturated soil in the voids.

Borehole reverberation and ringing noise signals were observed in the detection of a sinkhole. To address this, band-pass filtering was applied to remove high-frequency noise. Velocity estimation proves useful in distinguishing real diffraction from artifacts. The identification of a sinkhole necessitates consideration of the geological setting of the upper bedrock. If a karst void exists beneath a weathered and fractured upper bedrock layer, the enlargement of the borehole in the soil range may indicate soil loss and the development of a sinkhole.

When compared to seismic-based near-surface geophysical methods, GPR demonstrates higher sensitivity to water or saturated soil geological anomalies. Reflections from karst voids with water infill exhibit stronger signatures. EM waves decay more slowly in air, allowing for the detection range to remain unaffected by unfilled karst voids. Moreover, the single borehole GPR detection method offers time and cost savings due to the presence of pre-existing boreholes at the pile locations. Borehole-based detection provides enhanced accuracy and resolution as it is performed in closer proximity to the geological features. The proposed data processing method combining instantaneous phase analysis and medium filtering effectively extracts and highlights weak reflections from karst voids amidst borehole background noise in the near offset borehole radar detection.

Nevertheless, several limitations of the single borehole radar detection approach should be addressed in future studies. These include: (1) the absence of azimuthal information, which is crucial for remediation purposes but currently lacks sufficient research and application in the context of azimuthal single borehole radar; (2) challenges related to the use of PVC tubes to protect the borehole from collapse, particularly in cases where debris sediment at the bottom of the borehole and within karst voids hinders the smooth insertion of the PVC, thereby reducing the detection depth range; and (3) the need for more intuitive detection results comparable to cross-hole tomography, suggesting the potential adoption of full-waveform inversion for the inversion of single borehole radar data.

## References

Alperovich L, Eppelbaum L, Zheludev V et al. (2013) A new combined wavelet methodology applied to GPR and ERT data in the Montagnole experiment (French Alps). J Geophys Eng 10(2):025017, 1–17

Ebihara S (2004) Directional borehole radar with dipole antenna array using optical modulators. IEEE Trans Geosci Remote Sens 42(1):45–58

Li Y, L S, Liu B et al. (2017) Forward simulation and complex signal analysis of borehole radar detection for underground adverse geological bodies. Rock and Soil Mech 38(1):300–308

Liu B, Zhang F, Li S et al (2019) Full waveform inversion based on inequality constraint for cross-hole radar. J Appl Geophys 162:118–127

Ma C, Zhao Q, Ran L et al (2014) Numerical study of borehole radar for cliff imaging. J Environ Eng Geophys 19(4):269–276

Ma C, Zhao Q, Wang L et al (2015) Modeling borehole radar by finite-difference time-domain in conductive sandstone. J Environ Eng Geophys 20(1):19–29

Qin H, Vrugt J, Xie X et al (2018) Improved characterization of underground structure defects from two-stage Bayesian inversion using crosshole GPR data. Autom Constr 95:233–244

Taner M, Koehler F, Sheriff R (1979) Complex seismic trace analysis. Geophysics 44(6):1041–1063

Zhong S, Wang C, Wu L et al (2011) Borehole radar response characteristics of point unfavorable geo-bodies: forward simulation on its geometric effect. Rock and Soil Mech 32(5):1583–1588
Zhong S, Wang C, Wu L et al (2012) Borehole radar response characteristics of point unfavorable geo-bodies: Forward simulation of its surrounding rock and filling condition. Rock and Soil Mech 33(4):1191–1195

# Chapter 7
# Pile Hole Sonar Detection Equipment

## 7.1 Prologue

Although sonic detection of cavities around boreholes can enhance the investigation effectiveness of drilling, the detection depth is easily affected by sedimentation in the borehole. Moreover, when the advanced borehole is only slightly deeper than the design pile foundation, the inversion of sonic detection can be influenced by the bottom boundary of the borehole, making it difficult to cover the bottom part of the pile. Meanwhile, in areas with weak karst development, many engineering projects do not adopt the method of one hole per pile for investigation. However, since the presence of cavities at the bottom of the pile has a significant impact on the bearing capacity of the pile foundation, a missed detection of bottom cavities can easily lead to pile failure. Constructing the pile foundation without identifying the development of bottom cavities, can result in serious accidents such as cavity grouting failure, collapse, ground settlement, borehole burial, and cracks in surrounding buildings (Yan 2013). Therefore, it is necessary to propose a pile-bottom cavity detection method that combines the construction characteristics of cast-in-place piles, which is adaptable to engineering and can be widely applied for inspection.

Currently, there are several methods for detecting bottom cavities in pile foundations, as shown in Table 7.1 (Hu et al. 2009; Cui et al. 2011; Ge et al. 2004; Lian 2012; Zhang et al. 2014; Liu and Zhang 2014; Chen 2014; Zhao 2012): high-density electrical resistivity method, traditional seismic reflection method, ground-penetrating radar method, advanced drilling method, elastic wave CT method, and pipe wave testing method. However, due to the requirement of slurry wall protection for drilled cast-in-place piles, the high-density electrical resistivity method, traditional seismic reflection method, and ground-penetrating radar method are not commonly used for detecting cavities bottom piles, as they are difficult to operate in the slurry inside the pile holes. The elastic wave CT method and tube wave testing method can be used for detecting bottom cavities in drilled cast-in-place piles. However, the former requires


L. Liu et al., *Detection of Karst Voids at Deep Pile Foundation*,
https://doi.org/10.1007/978-981-97-5834-0_7

at least two boreholes around the pile during detection, and it has significant interference. The latter still requires advanced drilling for detecting bottom cavities and has difficulty in determining the orientation of the cavities. As a result, these methods have not been widely applied in engineering. In actual practice, the most common approach is to use one hole per pile, which is prone to missing cavities, while using multiple holes for each pile is too cumbersome. Therefore, during the construction and surveying phase, an effective and efficient detection method is needed to ensure that there are no cavities within a range of 3 times the pile diameter and not less than 5 m at the bottom of each pile [TNSCG 2009].

In Chapters 7 to 10 of this book, a sonar detection method for detecting cavities bottom drilled cast-in-place piles will be proposed. The aim is to explore the application of sonic technology in detecting bottom cavities by utilizing the slurry in the pile holes as the medium for sound wave propagation and coupling to investigate the development of cavities at the bottom of the piles. Sonic technology was invented in 1906 and was initially used in military applications. In the late-twentieth century, sonic technology began to transition to civilian use and found applications in underwater positioning and depth measurement (Tian 2011). In 1991, Zhong Shihang et al. (Zhong and Wang 2012; Zhong et al. 2008) first introduced the land-based sonic method in geological exploration, which is essentially an ultra-wideband elastic wave reflection method with extremely small offset distances. It achieved certain results in exploring caves and predicting tunnel geology on the ground and was mostly used in preliminary exploration stages. However, there have been limited research and application of this technique in construction surveys after pile hole drilling, and this technology has not been utilized in a liquid environment such as slurry in the pile holes. A novel sonar detection method for bottom cavities will be proposed in this chapter. A detection instrument for bottom cavity sonic detection (JL-SONAR) and signal analysis software (PBCA) will be designed. The intrinsic detection mechanism will be revealed through three-dimensional numerical simulations in Chap. 8, and effective signal analysis methods will be proposed. Finally, in Chap. 9, an automatic extraction method for effectively capturing reflected signals from bottom cavities will be presented.

## 7.2 Pile Hole Sonar Detection Scheme

The schematic diagram of pile hole sonar detection method for karst void beneath pile hole bottom, as designed in this paper, is shown in Fig. 7.1a. The detection system, including the pile hole probe and the on-site host, is depicted in Fig. 7.1b. Figure 7.1c illustrates the slow insertion of the sonar probe into the bottom of the pile during on-site detection. Figure 7.1d represents a one-transmitter-four-receiver pile hole sonar probe.

Detection of karst voids beneath pile hole can be divided into the following steps:

**Table 7.1** Comparison of detecting method for the karst cavity under pile bottom

| Method | Theory | Suitable for detecting karst cavity beneath a pile? | Additional drillings required | Detection extent | Detection precision | Whether it can used with slurry in the hole | Evaluation |
|---|---|---|---|---|---|---|---|
| Macro-geological analysis | Macro-analysis base on geological data | No | – | – | – | – | The prediction result is more macroscopic and cannot be used to make an accurate judgment |
| High-density apparent resistivity | Determine the conductive differences between the rocks and karst features | No | – | – | – | – | The infill of a karst cavity is complex. It is difficult to determine the corresponding conductivity |
| Traditional seismic reflection | Employs an artificial source to detect ground velocity variations | No | – | – | – | – | The bottom of the hole must be kept dry and smooth, which cannot be achieved in a bored pile |
| Ground-penetrating radar | Determine the dielectric differences among underground media | No | – | – | – | – | The transmitting and receiving antenna must be placed at the bottom of the pile hole, which is not possible in a bored pile |
| Transient electromagnetic method | Measure the subsequent decay response of induced electric and magnetic fields | No | – | – | – | – | The tests are easily disturbed by metal structures and cables |

(continued)

**Table 7.1** (continued)

| Method | Theory | Suitable for detecting karst cavity beneath a pile? | Additional drillings required | Detection extent | Detection precision | Whether it can used with slurry in the hole | Evaluation |
|---|---|---|---|---|---|---|---|
| Elastic wave CT | Subsurface tomographic imaging method | Yes | Need 2 ~ 5 drillings | Between the tested boreholes | More than 0.2 m | No | Two boreholes are needed to apply the CT method. The method cannot recognize karst features outside the profile between two test holes |
| Guided wave testing | Use tube waves to obtain information on the strata or lithology | Yes | Need 1 drilling | Approx. 1 m around the borehole | More than 0.3 m | No | The test depends on the advanced drillings at the pile position. The direction of a karst feature cannot be recognized |
| Advance drillings | Advance drillings at the pile bottom | Yes | Need 1 ~ 5 drillings | Within the drilled borehole | Based on the number of boreholes | No | Too few drillings could easily lead to a false negative. Excessive drillings for each pile make the geotechnical investigation too complex and time-consuming |

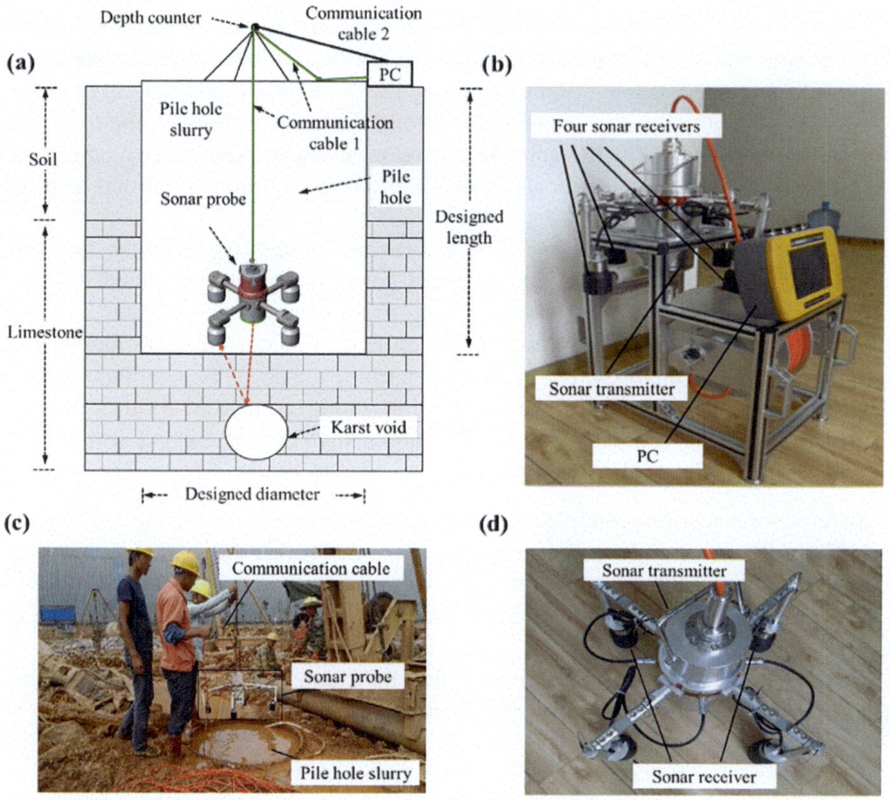


**Fig. 7.1** Pile hole sonar detection **a** schematic diagram, **b** detection system, **c** detection process photo, **d** pile hole sonar probe

(1) Probe Placement: The sonar detection probe is placed at the bottom of the pile using a communication cable and a hole mouth bracket. It is connected to the on-site host through the communication cable. If there is no slurry or water at the bottom of the artificial bore pile, 10–20 cm of water should be injected to ensure that the sonar transmitter and transducer of the probe are in contact with water, allowing the sonar stress waves to couple with the pile-bottom interface.
(2) Probe Leveling: The on-site PC utilizes the three-dimensional compass in the probe to obtain the attitude of the sonar detection probe at the pile bottom and the orientation of each sonar receiving transducer through the communication cable, ensuring that the sonar transmitter is nearly perpendicular to the bottom surface of the pile.
(3) Detection: Once the sonar detection probe is properly positioned, the on-site host controls the sonar transmission driver module and the sonar transmitter to emit sonar stress waves through the communication cable. It regulates the sonar signal processing module to receive the sonar signals, and the digitized data of

the received sonar are sent to the on-site host for display and processing via the communication cable and the communication port of the microcontroller.

(4) Multi-angle Detection: After completing the detection for the current data, the sonar transmitter can be rotated in a certain direction along with the receiving transducer, and Steps (2) and (3) can be repeated to conduct two additional detections, resulting in signals from 12 different directions in a total of three sets.

(5) Software Analysis: Using the PBCA software for karst voids sonar detection at the bottom of piles, all the received sonar signals are arranged in order of the orientation of the sonar receiving transducers. This generates a profile of the detected sonar stress waves, which is then subjected to comprehensive processing and analysis. The aim is to determine the development of karst voids at the bottom of the piles and ensure that within a range of 3 times the pile diameter and no less than 5 m, there are no karst voids or weak rock formations, as required (Cai et al. 2011).

In terms of detection principles, sonar, in a narrow sense, refers to the method and equipment that uses underwater sound waves to determine the presence, location, and type of objects. In a broader sense, any device or method that utilizes underwater sound waves as a propagation medium to achieve a certain purpose is referred to as sonar (Tian 2011). In this study, for the detection of karst voids at the bottom of drilled shafts, sonar stress waves generated in the slurry and water environment are used to detect the karst voids. The schematic diagram of the principle is shown in Fig. 7.2.

Transmission: The on-site host controls the sonar transmission driver module through the communication cable. The electrical signal is sent to the sonar transmitting transducer as shown in Fig. 5.2 and converted into sound waves, which are vertically emitted into the slurry environment at the bottom of the drilled shaft for karst void detection.

Propagation: Under the coupling effect of the slurry at the bottom of the shaft, the impedance difference between the upper and lower media is effectively reduced, increasing the sound coupling efficiency. This allows more sonar stress wave energy to smoothly enter the bedrock.

Reflection: After the sonar stress waves enter the bedrock and propagate downward, a distinct impedance boundary is formed between the karst voids, weak rock layers, and the rock mass at the bottom of the shaft. This is reflected in the variations of density and elastic wave propagation velocity in rocks or soil. As a result, this interface produces strong reflected echoes of the sonar stress waves.

Receiving: The reflected sonar waves are received by four sonar receiving transducers positioned around the sonar transmitting transducer. As shown in Fig. 7.2, the azimuth angles of the four receiving transducers are measured by a three-dimensional electronic compass installed inside the probe. The reflected sonar waves are converted into electrical signals by the receiving transducers and then amplified, filtered, and digitized. They are transmitted to the on-site host through the communication cable for display and processing.

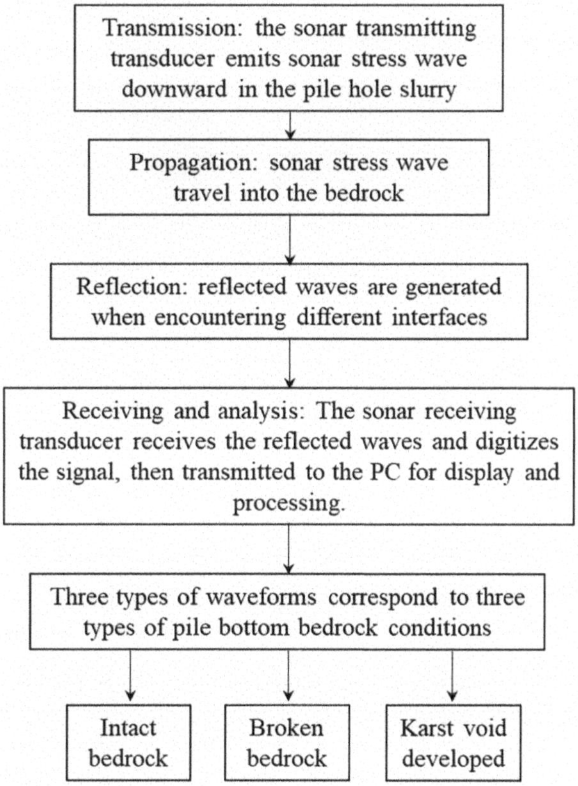


**Fig. 7.2** Pile hole sonar detection principle

## 7.3 Pile Hole Sonar Detection Equipment

1. (1) **Overall Scheme**

Based on the principles described in sect. 7.2, a set of sonar detection devices for karst voids at the bottom of drilled shafts has been developed in this study. It mainly consists of the pile hole Sonar Detection Instrument (JL-SONAR) and the Sonar Signal Analysis Software (PBCA). The developed bottom-hole Sonar Detection Instrument (JL-SONAR) is shown in Fig. 5.1b. It includes the on-site host placed at the surface near the pile borehole, the sonar detection probe positioned at the bottom of the pile borehole, and the communication cable connecting the host and the sonar detection probe. The structure of the Sonar Detection Instrument is illustrated in Fig. 7.3.

As shown in Fig. 7.3, the detection instrument mainly consists of the on-site host and the sonar detection probe. The on-site host comprises a central processor, human–machine interaction devices, and memory. The sonar detection probe includes a sonar transmitting transducer and sonar receiving transducers.

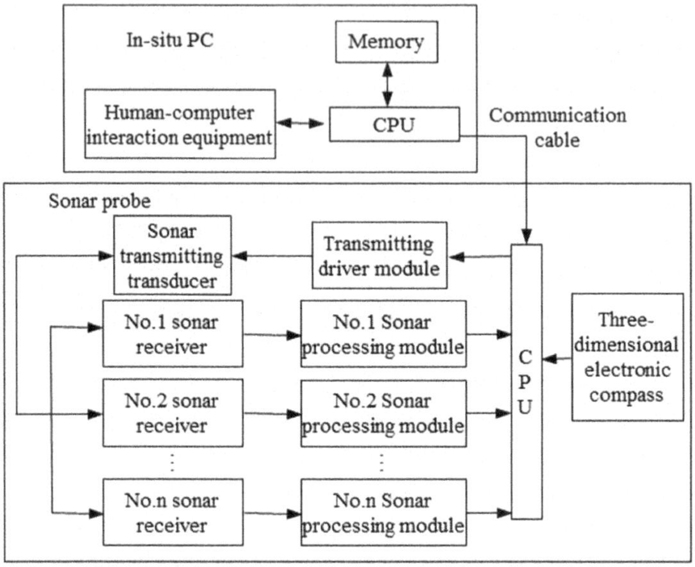


**Fig. 7.3** Structure of pile hole sonar detection equipment

The human–machine interaction devices in the host are responsible for inputting and outputting commands. The central processor primarily handles input instructions, generates corresponding operational control signals, guides the operation of the sonar detection probe, and performs calculations on the sonar detection data. The memory is used for storing the sonar detection data.

The sonar detection probe consists of a microcomputer, a three-dimensional electronic compass, a sonar transmitting device, and a sonar receiving device. The microcomputer controls and transfers instructions among the various components and facilitates information exchange with the on-site host. The three-dimensional electronic compass is horizontally installed inside the probe housing, with its north direction pointing toward the installation direction of the sonar receiving transducers. The sonar transmitting device consists of a sonar transmitting transducer and its corresponding driving module. The sonar receiving device includes sonar receiving transducers and a processing module, typically with four to eight sets of receiving devices placed around the transmitting device.

2. **Transmitting and Receiving Transducers**

The developed pile hole sonar probe is shown in Fig. 7.4, which includes a transmitting transducer located in the middle and four receiving transducers surrounding it. Similar to the karst cavity detection system introduced in Chap. 2, the transmitting transducer of the pile-bottom karst void detection system utilizes the same magnetostrictive principle and magnetostrictive material for low-frequency emission as the cavity detection system around the drilled hole. The receiving transducers also employ the same piezoelectric effect and piezoelectric ceramic material. Similar to the cavity detection system, a three-dimensional electronic compass is attached to one of the receiving transducers, allowing for the calculation of the azimuth angles of the other three transducers sequentially. The parameters of the sonar transmitting transducer (JL-5–1) and the sonar receiving transducer (JL-5–2) are shown in Tables 7.2 and 7.3, respectively.

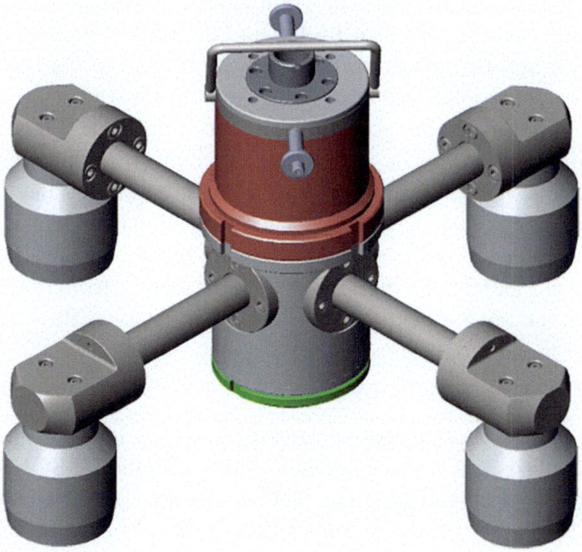

**Fig. 7.4** Pile hole sonar probe (one-transmitter-four-receiver)

**Table 7.2** Parameters of pile hole sonar transmitters

| Model type | Transmitting frequency | Maximum instantaneous power/Kw | Signal consistency | Sonar diffusion angle/degree |
|---|---|---|---|---|
| JL-5-1 | 200 Hz~8 kHz | 10 | > 99% | > 120 |
| Features | Working temperature/°C | Working mode | Pressure/MPa | Working voltage/V |
| High energy, broadband belt, short aftershock | −40~ + 80 | Single, continuous | 1.0 | 12 |

**Table 7.3** Parameters of pile hole sonar receiver

| Model type | Resonant frequency | Operating frequency | Capacitance | Sensitivity |
|---|---|---|---|---|
| JL-5-2 | 30 kHz ± 2 kHz | 100 Hz~10 kHz | 1 600 pF ± 20% | -190 dB |
| Minimum impedance | Working temperature/°C | Pressure/MPa | Shell material | Purpose |
| 400Ω ± 20% | −40~ + 80°C | 1.0 | Aluminum alloy | Sonar monitoring |

As shown in Table 7.1, the transmitting transducer reaches an emission power of 10 kW and operates in the frequency range of 200–8 kHz. The high frequency ensures high detection accuracy, while the combination of high-power and wide frequency band in the sonar emission prevents rapid attenuation of the high-frequency component, greatly enhancing signal recognizability.

During signal emission, the human–machine interaction device sends out a sonar stress wave transmission command. The command is transmitted to the microcomputer inside the sonar probe, and the microcomputer controls the sonar transducer through the sonar emission driver module to convert the electrical signal into an acoustic signal and emit the sonar stress wave.

During signal receiving, the four to eight sonar receiving transducers surrounding the emitting probe simultaneously convert the received acoustic signals into electrical signals. The signals then pass through the sonar signal processing module and enter the microcomputer. The sonar signal processing module consists of digital circuits and performs preprocessing on the received signals. Finally, the microcomputer gathers the digitally processed signals from multiple channels, transmits them to the on-site host for processing and analysis through communication cables, and completes data storage and output.

(1) Processing Software

The PBCA analysis software for the pile hole sonar detection device is a waveform analysis software designed to work with the JL-SONAR sonar detection device for pile-bottom cavities. The PBCA software is developed using Microsoft Visual Studio 2008 as the programming environment and utilizes C# as the programming language. It runs on Windows operating system and can be installed on the on-site host for real-time analysis or on a PC for post-analysis.

The structure of the PBCA software is shown in Fig. 7.5, and its main interface is depicted in Fig. 7.6. The PBCA signal analysis software consists of several modules, including the input module, waveform display and transformation module, signal processing module, annotation module, and output module.

The input module allows for the input of sonar signals, enabling integration of raw detection data files from different orientations of the same pile. It supports the simultaneous input of sonar signals from up to 12 different angle receiving transducers for the same pile.

The waveform display and transformation module arrange the sonar signals from different channels in the order of their azimuth angles to form stress wave profiles. It performs time-depth conversion and recovers the true amplitude of waveform data by exponentially amplifying the original waveform. It also allows for the comparison display of the original and processed waveforms for individual channel sonar signals.

The signal processing module provides various preliminary processing methods for the original waveforms:

(1) Frequency filtering: The signal is first Fourier transformed for spectral analysis to remove high-frequency and low-frequency interference waves. By selecting an appropriate filtering range, the analysis is focused on the frequency range of interest where the signal is truly valid.

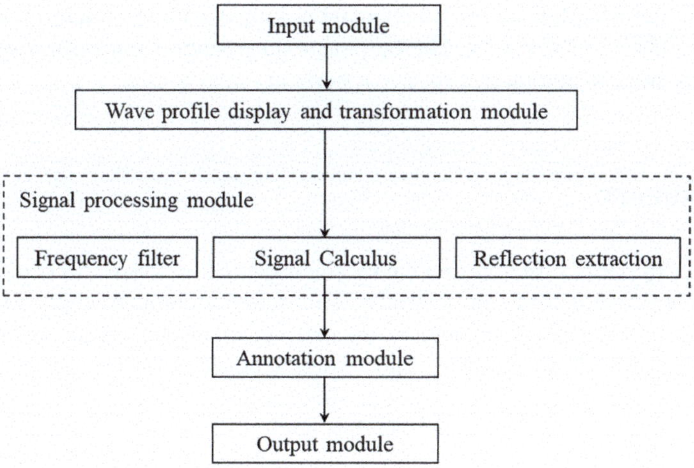


**Fig. 7.5** Structure of PBCA software

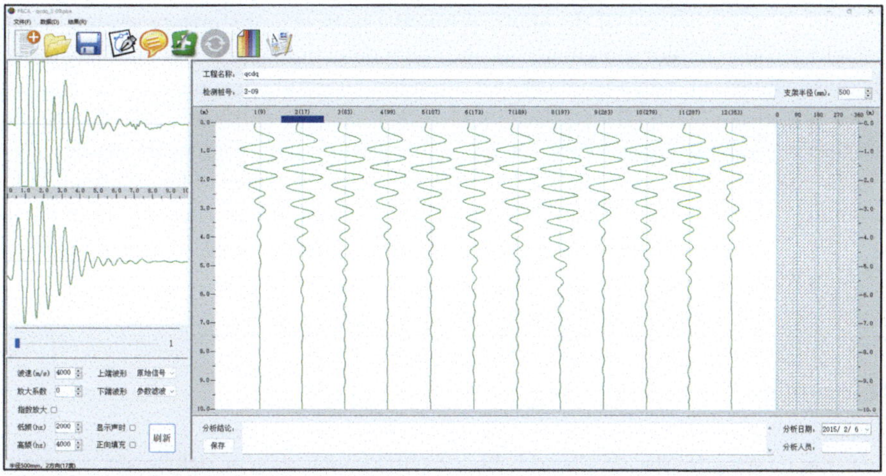

**Fig. 7.6** PBCA program interfaces

(2) Differentiation of signals: When reflections are not clearly visible, the signals are subjected to differentiation processing to amplify the degree of waveform distortion.
(3) Integration of signals: Integration processing is applied to the signals to reduce residual vibrations appropriately.
(4) Initial reflection extraction: Based on the phase difference of the same signal at different sampling times, suitable reflection coefficients are selected to clearly display the phase transition of the waveform.

The annotation module allows for manual marking of cavities, fractures, abnormal locations, shapes, or interfaces. The output module enables integration with other applications and can automatically save reports in Excel format.

## 7.4 Summary

This chapter presents a pile-bottom cavity sonar detection system for drilled shafts. The following conclusions can be drawn: This study attempts to apply sonar technology to the detection of cavities at the bottom of drilled shafts. It has developed a high-power, high-frequency underwater acoustic emission source and high-frequency underwater transducers, addressing the limitations of traditional reflection waves in adapting to the construction environment at the bottom of drilled shafts and achieving low cavity identification rates. Sonar technology effectively utilizes the slurry, which has acoustic impedance closer to the bedrock, as a coupling medium, reducing the impedance mismatch between different media and improving the acoustic coupling efficiency. This allows sound waves to penetrate smoothly into the bedrock at the bottom of the shaft. While the presence of slurry poses technical obstacles to other exploration methods, the sonar detection method ingeniously turns this disadvantage into an advantage. Additionally, the study designs the pile-bottom cavity detection device (JL-SONAR) and signal analysis software (PBCA), establishing a preliminary framework for rapid pile-bottom cavity sonar detection. The detection mechanism, wavefield characteristics, typical features of target anomalies, automatic extraction methods for reflection signals, and engineering applications of the drilled shaft pile-bottom cavity sonar detection method will be discussed in detail in Chaps. 8 and 9.

## References

Cai J, Yan C, Wang N, Shao Y, Zheng J, Tang Z (2011) Application of high-density resistivity method to karst investigation of metro engineering. J Eng Geol [in Chinese] 19(06):935–940

Chen J (2014) Research on the construction technology of the deep pile foundation in the strong karst development area. Master Thesis. Central South University, China

Cui Z, Xia C, Zhao K et al (2011) Discussion on the method and application of detection method for large diameter rock-socketed pile bottom cave. Western China Commun Sci and Technol [in Chinese]. 07(2011):46–50

Ge Z, Hu C, Xiang H (2004) The exploration of limestone karst cave at the end of pile. Geophys Geochem Explorat [in Chinese]. 01:85–87

Hu C, Deng S, Wang B (2009) Application study of the GPR method in the detection of karsts beneath artificial excavated hole pile. J Geol Hazards and Environ [in Chinese] 20(02):74–78

Lian G (2012) Research on the detection technology of cave on the bottom of the cast-in-place pile. Master. Thesis. Chongqing Jiaotong University, China

Liu W, Zhang Z (2014) The application of ground-penetrating radar to the karst foundation investigation. Geophys Geochem Explorat [in Chinese]. 03:624–628

The National Standards Compilation Group of People's Republic of China (2009) GB50021—2009 Code for investigation of geotechnical engineering. China Architecture and Building Press, Beijing, pp 52–55

Tian T (2011) Sonar Technology. Harbin Engineering University Press

Yan S (2013) Research on the stability of cave roof under pile foundation. Master Thesis. Central South University, China

Zhang Z, Sun T, Liu J (2014) Frequency domain analysis of ground penetrating radar in detecting the karst rock cavity under pile tips. Soil Eng Foundation [in Chinese]. 02:163–166

Zhao J (2012) Research of tube wave prospecting karst development under large diameter pile. Master Thesis. Ocean University of China, China

Zhong S, Wang R (2012) The progress of geophysical technologies for the downtown area of a city and the breakthrough in karst cave survey and rock burst prediction of the tunnel. Geophys Geochem Explorat [in Chinese]. 36(03):511–516

Zhong S, Sun H, Wang R, Li S, Ran M (2007) The superiority of land sonar in karst cave survey. Geophys Geochem Explorat [in Chinese]. S1:111–115

# Chapter 8
# Pile Hole Sonar Detection Analysis Method

## 8.1 Prologue

Chapter 7 presents a method for detecting the development of bottom dissolution cavities during the construction phase using pile foundation borehole drilling, which is defined as the bottom dissolution cavity sonar detection method. This method is stimulated in a liquid slurry environment and fully utilizes the impedance difference between the slurry in the pile hole and the bedrock, thereby achieving coupling. Based on this, Chap. 7 designs the bottom dissolution cavity sonar detection equipment. However, due to the complex development of bottom dissolution cavities, the detected signals are often very complex. It is highly challenging to interpret the detection mechanism, deepen the understanding of the detection process, and propose effective signal analysis methods based on this.

Numerical simulation is an effective means to understand the propagation, reflection, and detection influencing factors of sonar stress. At the same time, numerical simulation can provide a basis for improving detection hardware and proposing signal analysis methods. In this chapter, starting from the three-dimensional numerical simulation of bottom dissolution cavity sonar detection, the three-dimensional crossed-grid finite-difference method in the time domain is used to numerically simulate the detection process and detection signals, revealing the detection mechanism. Based on the detection mechanism, qualitative fast analysis and single-channel signal analysis methods based on the generalized S-transform are proposed, and numerical signals and on-site signals are used to validate the signal processing methods.



L. Liu et al., *Detection of Karst Voids at Deep Pile Foundation*,
https://doi.org/10.1007/978-981-97-5834-0_8

## 8.2 Three-Dimensional Numerical Simulation Research

In this section, we will create a model of pile hole-bedrock-karst voids using field detection data. Numerical simulations will be conducted using the three-dimensional viscoelastic wave equation and the three-dimensional time-domain crossed-grid finite-difference method. The simulation parameters will be adjusted using field data to ensure the accuracy of the results. The test conditions will cover various dissolution cavities with different depths, sizes, and orientations, as well as factors that influence detection, including receiver positions, pile diameter, waveform frequency, and cavity fillings.

### *8.2.1 Numerical Method*

A set of 3D viscoelastic wave equations, which account for energy attenuation, is employed to simulate sonar detection and visualize karst cavities beneath bored cast-in-situ piles. The high-order staggered-grid finite-difference (SGFD) method is utilized to discretize and solve these wave equations within a Cartesian coordinate system. Perfectly matched layer (PML) boundary conditions are implemented to minimize reflections from the computational domain's edges. The following sections provide a comprehensive explanation of the 3D wave equations, the SGFD method, and the boundary, initial, and stability conditions involved in the modeling process.

(1) **Three-Dimensional Viscoelastic Wave Equations**

Equations (8.1–8.9) show the 3D first-order stress and velocity viscoelastic wave equations in isotropic media (Saenger and Bohlen 2002):

$$\frac{\partial v_x}{\partial t} = \frac{1}{\rho}\left(\frac{\partial \tau_{xx}}{\partial x} + \frac{\partial \tau_{xy}}{\partial y} + \frac{\partial \tau_{xz}}{\partial z}\right), \tag{8.1}$$

$$\frac{\partial v_y}{\partial t} = \frac{1}{\rho}\left(\frac{\partial \tau_{xy}}{\partial x} + \frac{\partial \tau_{yy}}{\partial y} + \frac{\partial \tau_{yz}}{\partial z}\right), \tag{8.2}$$

$$\frac{\partial v_z}{\partial t} = \frac{1}{\rho}\left(\frac{\partial \tau_{xz}}{\partial x} + \frac{\partial \tau_{yz}}{\partial y} + \frac{\partial \tau_{zz}}{\partial z}\right), \tag{8.3}$$

$$\frac{\partial \tau_{xx}}{\partial t} = (\lambda + 2G)\frac{\partial v_x}{\partial x} + \lambda\left(\frac{\partial v_y}{\partial y} + \frac{\partial v_z}{\partial z}\right) - \varepsilon_L \tau_{xx}, \tag{8.4}$$

$$\frac{\partial \tau_{yy}}{\partial t} = (\lambda + 2G)\frac{\partial v_y}{\partial y} + \lambda\left(\frac{\partial v_x}{\partial x} + \frac{\partial v_z}{\partial z}\right) - \varepsilon_L \tau_{yy}, \tag{8.5}$$

$$\frac{\partial \tau_{zz}}{\partial t} = (\lambda + 2G)\frac{\partial v_z}{\partial z} + \lambda\left(\frac{\partial v_x}{\partial x} + \frac{\partial v_y}{\partial z}\right) - \varepsilon_L \tau_{zz}, \tag{8.6}$$

$$\frac{\partial \tau_{xy}}{\partial t} = G\left(\frac{\partial v_x}{\partial y} + \frac{\partial v_y}{\partial x}\right) - \varepsilon_T \tau_{xy}, \tag{8.7}$$

$$\frac{\partial \tau_{xz}}{\partial t} = G\left(\frac{\partial v_x}{\partial z} + \frac{\partial v_z}{\partial x}\right) - \varepsilon_T \tau_{xz}, \tag{8.8}$$

$$\frac{\partial \tau_{yz}}{\partial t} = G\left(\frac{\partial v_y}{\partial z} + \frac{\partial v_z}{\partial y}\right) - \varepsilon_T \tau_{yz}. \tag{8.9}$$

The equations of motion are represented by the particle velocity vectors $v_x$, $v_y$, and $v_z$, while the particle stress vectors are denoted as $\tau_{xx}$, $\tau_{xy}$, $\tau_{xz}$, $\tau_{yy}$, $\tau_{yz}$, and $\tau_{zz}$. These quantities are influenced by the density $\rho$ of the medium, as well as the Lamé parameters $\lambda$ and $G$. The Lamé parameters can be calculated using the P-wave velocity $v_p$ and the S-wave velocity $v_s$. To account for energy attenuation in the slurry environment, the impedance coefficients $\varepsilon_L$ and $\varepsilon_T$ are introduced to represent the impedance of P-waves and S-waves, respectively (Hosokawa 2010, 2011). The values of these impedance coefficients are determined through calibration based on experimental test signals.

(2) **Initial Conditions**

The model system is initially in a static state; therefore, the velocity and stress vectors are all zero. A Ricker wavelet, depicted in Fig. 8.1, is then emitted from the central point of the pile hole bottom, propagating through the system as follows:

$$\tau_{xx}(x, y, z, t) = A \times \left(1 - 2(\pi f_m t)^2\right) \exp\left[-(\pi f_m t)^2\right]. \tag{8.10}$$

The wavelet is transmitted from the center point $(x, y, z)$ of the pile hole bottom. It has a period of t, a dominant frequency of $f_{\mathrm{m}}$, and an amplitude denoted by $A$.

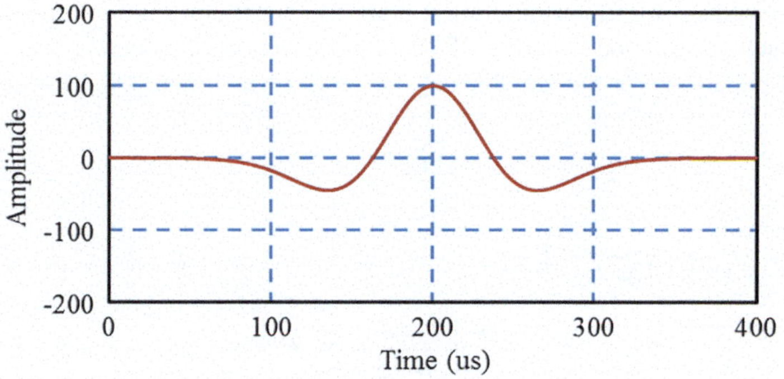


**Fig. 8.1** Ricker wavelet used in the simulation has a dominant frequency of 3 kHz and an amplitude of 100

### (3) High-Order Difference Equation Using PML Boundary Conditions

To solve the 3D viscoelastic wave equations, the high-order staggered-grid finite-difference (SGFD) method is applied in a Cartesian coordinate system. To mitigate numerical dispersion, a high-order spatial accuracy is utilized (Dong et al. 2000b). In this book, a tenth-order accuracy in the spatial domain ($2N = 10$) is employed, coupled with a second-order accuracy in the time domain. This level of accuracy surpasses the typical second-order accuracy employed in previous geophysical studies (Lu et al. 2013), ensuring precise modeling for sonar detection purposes.

To minimize reflections from the boundaries of the computational domain, perfectly matched layers (PMLs) are commonly employed (Collino and Tsogka 2001). PMLs act as absorbing layers that have a wave impedance matching that of the adjacent medium. This enables rapid decay and attenuation of waves propagating through the PML, effectively reducing the impact of boundary reflections. The PMLs are applied around the computational domain, dividing the three-dimensional components $(x, y, z)$ of the wave based on the PML conditions at the boundary region. Damping factors $(d_x, d_y, d_z)$ are utilized in the respective directions (Marcinkovich and Olsen 2003) as outlined below:

$$d_x = \frac{3}{2}\log\left(\frac{1}{R}\right)V\frac{x^2}{\delta^3}_{\max}, \tag{8.11}$$

$$d_y = \frac{3}{2}\log\left(\frac{1}{R}\right)V\frac{y^2}{\delta^3}_{\max}, \tag{8.12}$$

$$d_z = \frac{3}{2}\log\left(\frac{1}{R}\right)V\frac{z^2}{\delta^3}_{\max}, \tag{8.13}$$

where $R$ is the theoretical reflection coefficient, $V_{\max}$ is the maximum wave velocity, $\delta$ is the PML thickness, and $d_x = d_y = d_z = 0$ when $(x, y, z)$ is within the computational domain. Based on the aforementioned high-order staggered-grid finite-difference (SGFD) method and the described PML conditions, we can derive the following difference equations for the variables $v_x$, $\tau_{xx}$, and $\tau_{xy}$ as illustrative examples:

$$v_x = v_x^x + v_x^y + v_x^z, \tag{8.14}$$

$$v_{x\,i,j,k}^{x\,t+1/2} = \left(\frac{1}{1 + 0.5\Delta t \cdot d_x}\right) \cdot \left[(1 - 0.5\Delta t \cdot d_x)v_{x\,i,j,k}^{x\,t-\frac{1}{2}} + \frac{\Delta t}{\rho \cdot \Delta x}\sum_{n=1}^{N} C_n^{(N)}\left(\tau_{xx}{}^{t}_{i+\frac{2n-1}{2},j,k} - \tau_{xx}{}^{t}_{i-\frac{2n-1}{2},j,k}\right)\right], \tag{8.15}$$

$$v_{x\,i,j,k}^{y\,t+1/2} = \left(\frac{1}{1+0.5\Delta t\cdot d_y}\right)\cdot\left[(1-0.5\Delta t\cdot d_y)v_{x\,i,j,k}^{y\,t-\frac{1}{2}} + \frac{\Delta t}{\rho\cdot\Delta y}\sum_{n=1}^{N} C_n^{(N)}\left(\tau_{xy^t_{i,j+\frac{2n-1}{2},k}} - \tau_{xy^t_{i,j-\frac{2n-1}{2},k}}\right)\right], \tag{8.16}$$

$$v_{x_{i,j,k}}^{z_{t+1/2}} = \left(\frac{1}{1+0.5\Delta t\cdot d_z}\right)\cdot\left[(1-0.5\Delta t\cdot d_z)v_{x^{i,j,k}}^{z^{t-\frac{1}{2}}} + \frac{\Delta t}{\rho\cdot\Delta z}\sum_{n=1}^{N} C_n^{(N)}\left(\tau_{xz^t_{i,j,k+\frac{2n-1}{2}}} - \tau_{xz^t_{i,j,k-\frac{2n-1}{2}}}\right)\right], \tag{8.17}$$

$$\tau_{xx} = \tau_{xx}^{x} + \tau_{xx}^{y} + \tau_{xx}^{z}, \tag{8.18}$$

$$\tau_{xx_{i+1/2,j+1/2k i+1/2}}^{x^{t+1}} = \left(\frac{1}{1+0.5\Delta t\cdot d_x}\right)\left[(1-0.5\Delta t\cdot d_x)\tau_{xx\,i+\frac{1}{2},j+\frac{1}{2},k+\frac{1}{2}}^{x\,t+\frac{1}{2}} + \frac{(\lambda+2G)\Delta t}{\Delta x}\sum_{n=1}^{N} C_n^{(N)}\left(v_{x_{i+\frac{1}{2}+\frac{2n-1}{2},j+\frac{1}{2},k+\frac{1}{2}}}^{t+\frac{1}{2}} - v_{x_{i+\frac{1}{2}-\frac{2n-1}{2},j+\frac{1}{2},k+\frac{1}{2}}}^{t+\frac{1}{2}}\right)\right] - \varepsilon_L\tau_{xx\,i+\frac{1}{2},j+\frac{1}{2},k+\frac{1}{2}}^{x\,t+\frac{1}{2}} \tag{8.19}$$

$$\tau_{xx\,i+1/2,j+1/2,k+1/2}^{y\,t+1} = \left(\frac{1}{1+0.5\Delta t\cdot d_y}\right)\left[(1-0.5\Delta t\cdot d_y)\tau_{xx\,i+\frac{1}{2},j+\frac{1}{2},k+\frac{1}{2}}^{y\,t+\frac{1}{2}} + \frac{\lambda\Delta t}{\Delta y}\sum_{n=1}^{N} C_n^{(N)}\left(v_{y_{i+\frac{1}{2},j+\frac{1}{2}+\frac{2n-1}{2},k+\frac{1}{2}}}^{t+\frac{1}{2}} - v_{y_{i+\frac{1}{2},j+\frac{1}{2}+\frac{2n-1}{2},k+\frac{1}{2}}}^{t+\frac{1}{2}}\right)\right] - \varepsilon_L\tau_{xx\,i+\frac{1}{2},j+\frac{1}{2},k+\frac{1}{2}}^{y\,t+\frac{1}{2}}, \tag{8.20}$$

$$\tau_{xx\,i+1/2,j+1/2,k+1/2}^{z\,t+1} = \left(\frac{1}{1+0.5\Delta t\cdot d_z}\right)\left[(1-0.5\Delta t\cdot d_z)\tau_{xx\,i+\frac{1}{2},j+\frac{1}{2},k+\frac{1}{2}}^{z\,t+\frac{1}{2}} + \frac{\lambda\Delta t}{\Delta z}\sum_{n=1}^{N} C_n^{(N)}\left(v_{z_{i+\frac{1}{2},j+\frac{1}{2},k+\frac{1}{2}+\frac{2n-1}{2}}}^{t+\frac{1}{2}} - v_{z_{i+\frac{1}{2},j+\frac{1}{2},k+\frac{1}{2}+\frac{2n-1}{2}}}^{t+\frac{1}{2}}\right)\right]$$

$$- \varepsilon_L \tau^{z\,t+\frac{1}{2}}_{xx\,i+\frac{1}{2},j+\frac{1}{2},k+\frac{1}{2}}, \tag{8.21}$$

$$\tau_{xy} = \tau^x_{xy} + \tau^y_{xy}, \tag{8.22}$$

$$\begin{aligned}\tau^{x\,t+1}_{xy\,i+1/2,j+1/2,k+1/2} = {} & \left(\frac{1}{1+0.5\Delta t \cdot d_x}\right) \\ & \left[(1-0.5\Delta t \cdot d_x)\tau^{x^{t+\frac{1}{2}}}_{xy_{i+\frac{1}{2},j+\frac{1}{2},k+\frac{1}{2}}} + \frac{G\Delta t}{\Delta x}\right. \\ & \left.\sum_{n=1}^{N} C_n^{(N)} \left( v_{y^{t+\frac{1}{2}}_{i+\frac{1}{2}+\frac{2n-1}{2},j+\frac{1}{2},k+\frac{1}{2}}} - v_{y^{t+\frac{1}{2}}_{i+\frac{1}{2}+\frac{2n-1}{2},j+\frac{1}{2},k+\frac{1}{2}}} \right)\right] \\ & - \varepsilon_T \tau^{x\,t+\frac{1}{2}}_{xy\,i+\frac{1}{2},j+\frac{1}{2},k+\frac{1}{2}},\end{aligned} \tag{8.23}$$

$$\begin{aligned}\tau^{y\,t+1}_{xy\,i+1/2,j+1/2,k+1/2} = {} & \left(\frac{1}{1+0.5\Delta t \cdot d_y}\right) \\ & \left[(1-0.5\Delta t \cdot d_y)\tau^{y\,t+\frac{1}{2}}_{xy\,i+\frac{1}{2},j+\frac{1}{2},k+\frac{1}{2}} + \frac{G\Delta t}{\Delta y}\right. \\ & \left.\sum_{n=1}^{N} C_n^{(N)} \left( v_{x^{t+\frac{1}{2}}_{i+\frac{1}{2},j+\frac{1}{2}+\frac{2n-1}{2},k+\frac{1}{2}}} - v_{x^{t+\frac{1}{2}}_{i+\frac{1}{2},j+\frac{1}{2}-\frac{2n-1}{2},k+\frac{1}{2}}} \right)\right] \\ & - \varepsilon_T \tau^{y\,t+\frac{1}{2}}_{xy\,i+\frac{1}{2},j+\frac{1}{2},k+\frac{1}{2}},\end{aligned} \tag{8.24}$$

where $\Delta t$ is the time step (0.002 ms in this study) and $C_n{}^{(N)}$ is the difference coefficient. Here, $C_1{}^{(5)} = 1.211243$, $C_2{}^{(5)} = -0.08972168$, $C_3{}^{(5)} = 0.001384277$, $C_4{}^{(5)} = -0.00176566$, and $C_5{}^{(5)} = 0.0001186795$.

(4) **Continuity Conditions**

To maintain velocity and stress continuity on the material cell surfaces within an isotropic heterogeneous medium, a suitable discretization of the material parameters ρ and G is required. A material parameter grid is employed to align with the integration cells for τ. Specifically, we define ρ to calculate $vx$, $vy$, and $vz$, and we define G to compute $\tau xy$, $\tau xz$, and $\tau yz$, ensuring that the shear stress is zero at the solid–liquid interface (Schubert et al. 1998).

$$\rho\left(i+\frac{1}{2},j,k\right) = \frac{\rho(i+1,j,k)+\rho(i,j,k)}{2}, \tag{8.25}$$

$$\rho\left(i,j+\frac{1}{2},k\right) = \frac{\rho(i,j+1,k)+\rho(i,j,k)}{2}, \tag{8.26}$$

$$\rho\left(i,j,k+\frac{1}{2}\right) = \frac{\rho(i,j,k+1)+\rho(i,j,k)}{2}, \tag{8.27}$$

$$\frac{4}{G_{i+\frac{1}{2},j+\frac{1}{2},k}} = \frac{1}{G_{i,j,k}} + \frac{1}{G_{i+1,j,k}} + \frac{1}{G_{i,j+1,k}} + \frac{1}{G_{i+1,j+1,k}}, \tag{8.28}$$

$$\frac{4}{G_{i+\frac{1}{2},j,k+\frac{1}{2}}} = \frac{1}{G_{i,j,k}} + \frac{1}{G_{i+1,j,k}} + \frac{1}{G_{i,j,k+1}} + \frac{1}{G_{i+1,j,k+1}}, \tag{8.29}$$

$$\frac{4}{G_{i,j+\frac{1}{2},k+\frac{1}{2}}} = \frac{1}{G_{i,j,k}} + \frac{1}{G_{i,j+1,k}} + \frac{1}{G_{i,j,k+1}} + \frac{1}{G_{i,j+1,k+1}}. \tag{8.30}$$

(5) **Stability Condition**

In order to maintain computational stability, certain conditions must be satisfied regarding the spatial and temporal sampling intervals, as outlined by (Dong et al. 2000a).

$$v_{\max}\Delta t\sqrt{\left(\frac{1}{\Delta x}\right)^2 + \left(\frac{1}{\Delta y}\right)^2 + \left(\frac{1}{\Delta z}\right)^2} \le 1, \tag{8.31}$$

where $\Delta x$, $\Delta y$, and $\Delta z$ are the spatial sampling intervals (0.05 m in this study). The stability conditions are uniform and not dependent on Poisson's ratio, as the velocity of an elastic wave remains constant regardless of the direction within a 3D isotropic elastic medium. Consequently, the 3D staggered-grid finite-difference (SGFD) method can effectively model the propagation of elastic waves in media with varying Poisson ratios while maintaining stability.

## *8.2.2 Numerical Model and Verification*

(1) **Numerical Model**

Figure 8.2 illustrates a 3D computational model depicting a bored pile hole situated above a karst cavity embedded in the bedrock. The dimensions of the computational domain measure 8 × 8 × 8 m. The pile hole is positioned at the upper center of the model, featuring a diameter of 1.6 m and a depth of 2 m. A vertically incident wave is transmitted from the central point of the hole bottom, with the transmitter coordinates being (0, 0, 0).

To minimize reflections from the artificial boundaries, three distinct regions of perfectly matched layers (PMLs) are incorporated in the computational domain, as depicted in Fig. 8.2. Each PML region has a thickness of 2 m. In boundary area 1, located at the eight corners of the domain and marked in red, the damping factors dx, dy, and dz are set to nonzero values. In boundary area 2, denoted by the blue regions along the edges of the domain, let's consider the edge formed by the intersection of the xoy and yoz planes as an example. Here, d$x$ and d$z$ are set to nonzero values. In boundary area 3, represented by the green areas, d$x$ is nonzero along the two planes

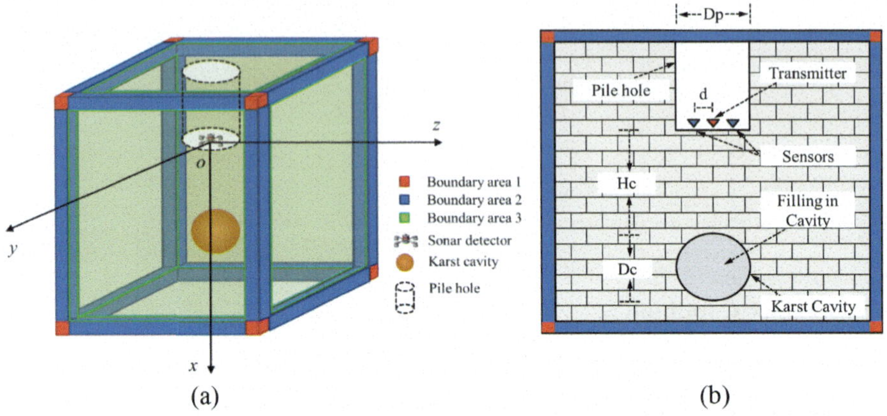


**Fig. 8.2** Numerical model of the detection of a karst cavity under a bored pile: **a** three-dimensional view and **b** two-dimensional view

perpendicular to the x-axis. Similar methods are applied to determine the damping factors in other planes.

Thirteen different scenarios are explored to examine the wavefield dynamics during the detection process and analyze the characteristics of karst cavities, including their depth, size, and orientation. Additionally, the impact of factors such as the optimal receiver position, pile hole diameter, and transmitted source frequency is investigated. The material properties of the various media involved in the study are detailed in Table 8.1. The geometrical and detection parameters for each numerical modeling scenario are presented in Table 8.2.

The modeling process was performed using the MATLAB software environment and involved the following steps: (1) Construction of the SGFD model: The model was built based on the pile hole and the geological structure beneath it. Appropriate spatial and temporal resolutions were chosen to balance precision and computational cost. The total number of spatial grids utilized was 13,824,000 (240 × 240 × 240), which comprised the core computational domain with 4,096,000 grids (160 × 160 × 160) and the boundary domain with 9,728,000 grids. (2) Input of parameters: All relevant parameters, such as the transmitter source frequency and the positions of the transmitter source and sensors, were provided as input. (3) Input of difference coefficients: The difference coefficients, ensuring second-order accuracy in the time domain and tenth-order accuracy in the spatial domain, were specified. (4) Input of PML boundary conditions: Three types of boundary conditions were established

**Table 8.1** Material properties of the media in the numerical model

| Material properties | $v_p$ (m/s) | $v_s$ (m/s) | $\rho$ (kg/m$^3$) |
|---|---|---|---|
| Bedrock | 4000 | 2300 | 2300 |
| Slurry in the pile hole | 2000 | 0 | 1500 |
| Cavity infill | 2000 | 1100 | 1500 |

Table 8.2 Geometrical and detection parameters of the model scenarios

| Scenario | Depth of cavity roof Hc (m) | Diameter of cavity Dc (m) | Coordinate of cavity center (m) | Diameter of pile hole (m) | Frequency of transmitted source (kHz) | Cavity infill |
|---|---|---|---|---|---|---|
| 1 | – | – | – | 1.6 | 3 | – |
| 2 | 1 | 2 | (2, 0, 0) | 1.6 | 3 | Clay |
| 3 | 3 | 2 | (4, 0, 0) | 1.6 | 3 | Clay |
| 4 | 5 | 2 | (6, 0, 0) | 1.6 | 3 | Clay |
| 5 | 3 | 1 | (3.5, 0, 0) | 1.6 | 3 | Clay |
| 6 | 3 | 3 | (4.5, 0, 0) | 1.6 | 3 | Clay |
| 7 | 3 | 2 | (4, 2, 1) | 1.6 | 3 | Clay |
| 8 | 3 | 2 | (4, 0, 0) | 1.2 | 3 | Clay |
| 9 | 3 | 2 | (4, 0, 0) | 2.0 | 3 | Clay |
| 10 | 3 | 2 | (4, 0, 0) | 1.6 | 2 | Clay |
| 11 | 3 | 2 | (4, 0, 0) | 1.6 | 5 | Clay |
| 12 | 3 | 2 | (4, 0, 0) | 1.6 | 3 | Water |
| 13 | 3 | 2 | (4, 0, 0) | 1.6 | 3 | Air |

based on the surrounding computational domain, following the PML principles. (5) Simulation of wave propagation: The Finite-Difference Time-Domain (FDTD) method was employed to simulate wave propagation. The stresses and velocities in space, encompassing both the computational domain and the PML condition domain, were calculated using Eqs. 8.14–8.24 throughout the simulation period. It should be noted that the initial stresses and velocities were set to zero. (6) Reception and analysis of the test signal: The sensor at the bottom of the pile received and analyzed snapshots, wave profiles, and test signals to evaluate the results of the simulation.

### (2) **Calibration and Verification of the Numerical Model**

To validate the modeling approach and calibrate wave attenuation through the slurry, two field detection experiments were conducted, as depicted in Fig. 8.3. These experiments took place as part of the Ping-Wen expressway project, specifically the Pingyuan to Laohuilong section, located in southeastern Yunnan Province, China. The expressway project spanned a total length of 33.814 km and included 34 bridges, covering a distance of 5.07 km. The geological conditions in the area were influenced by a major fault with a northwesterly trend, leading to the formation of a graben basin. The presence of multiple faults near the construction area resulted in the fracturing of carbonate rocks, creating favorable conditions for the development of karst cavities. The karst formations in the region consist of thick layers of carbonate rocks. Additionally, Quaternary deposits, including clays and gravels, are present with varying thicknesses ranging from a few meters to tens of meters.

During the field detection experiments, a JL-SONAR device (Shi et al. 2016) was utilized to capture the test signal with a shot interval (d) of 0.4 m. The experimental parameters were consistent with those of Scenarios 1 and 3. In Scenario 1, the bedrock

**Fig. 8.3** Field tests in a pile hole with a diameter of 1.6 m

beneath the test hole comprised intact limestone, while in the other scenario, a karst cavity filled with clay was present at a depth of 3 m.

In Fig. 8.4, the red curve represents the experimental signal obtained without a karst cavity, while the dotted line corresponds to the modeled signal for Scenario 1 with the impedance coefficients $\varepsilon_L$ and $\varepsilon_T$ set to zero. The decay of the elastic wave energy propagating through the slurry, due to its porous structure, necessitates the calibration of the impedance coefficients. After calibration, the modeled signal with $\varepsilon_L = 5000\ \text{s}^{-1}$ and $\varepsilon_T = 7000\ \text{s}^{-1}$ in the slurry (black curve in Fig. 8.4) exhibits relatively good agreement with the experimental signal. However, slight discrepancies between the modeled and experimental data may arise due to factors such as detector effects on wave propagation and variations between the actual transmitted wavelet and the Ricker wavelet.

Figure 8.5 showcases the experimental and modeled signals from Scenario 3. Points A and B in the figure correspond to the starting points of the reflected signals from the cavity roof and floor, respectively. Overall, the modeled signal, incorporating calibrated values of εL and εT (black curve in Fig. 8.7), aligns well with the test signal (red curve in Fig. 8.5), confirming the feasibility of the numerical modeling approach and the applicability of the impedance coefficients for the slurry. Minor disparities between the modeled and experimental signals can be attributed to the complex viscoelastic behavior of the karst cavity infill.

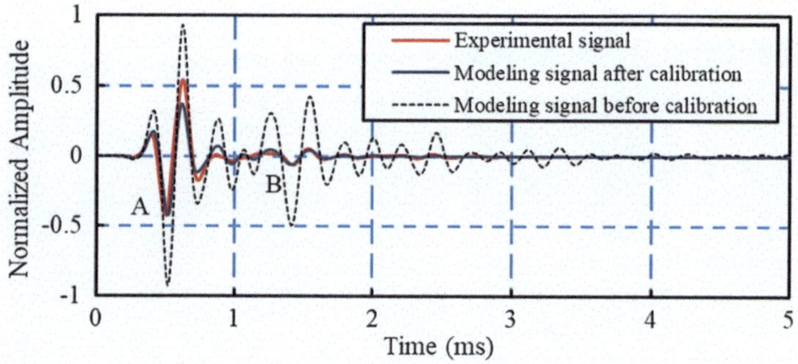


**Fig. 8.4** Comparison of the experimental and modeled signals for intact bedrock without a karst cavity (Scenario 1)

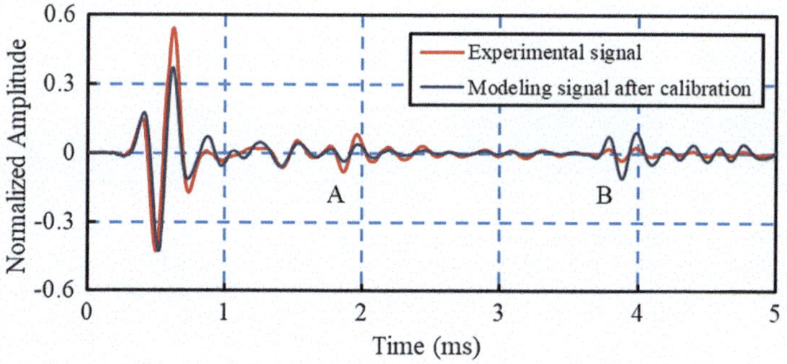


**Fig. 8.5** Comparison of the experimental and modeled signals for a karst cavity at depth of 3 m (Scenario 3)

### *8.2.3 Numerical Analysis*

(1) **Analysis of the Wavefield**

To visualize the wavefield during the sonar detection of a karst cavity beneath a bored pile in Scenario 3 (as specified in Table 8.2), snapshots of the modeling results were generated. Figure 8.6 displays snapshots of the vertical velocity in the xoz plane, which intersects the center of the pile hole bottom.

At a specific time of 0.2 ms (as depicted in Fig. 8.6), the waves transmitted from the center of the pile hole bottom can be categorized into three types. Firstly, there are waves traveling upwards into the slurry. Secondly, waves propagate horizontally along the hole bottom. Lastly, waves travel downwards into the bedrock. Among these three types of waves, the wave traveling upwards comprises a direct transmitted wave and a wave reflected from the hole bottom. However, since this wave will not reach the sensor within the relatively short detection time, it can be disregarded and excluded from the analysis. Additionally, the PML boundary employed in the

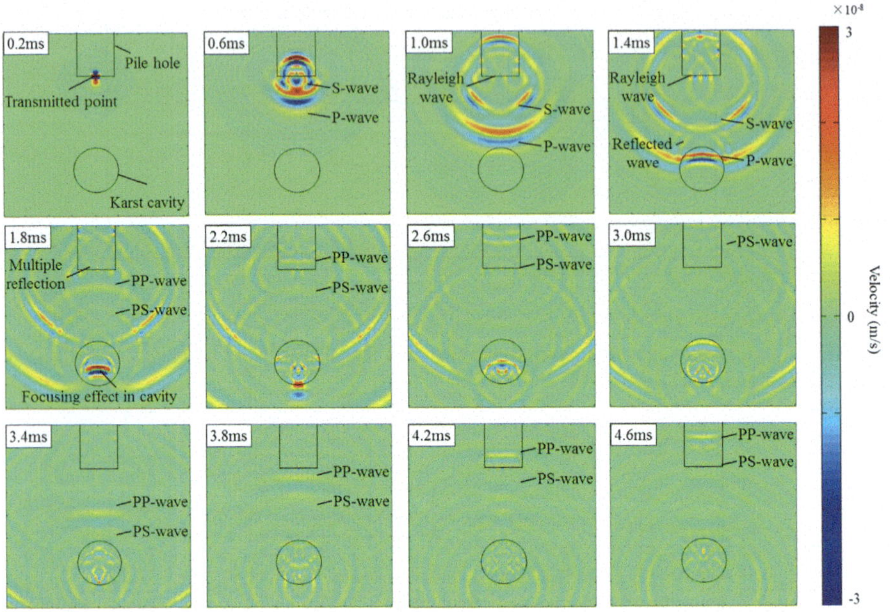


**Fig. 8.6** Snapshots of the vertical velocity in the *xoz* plane intersecting the center of the pile hole bottom (the snapshots here are based on Scenario 3). The PP-wave and PS-wave are characterized by their propagation as P-waves into the bedrock and subsequent reflection as P-waves and S-waves, respectively

numerical study effectively eliminates this wave. Therefore, the subsequent wave propagation analysis will concentrate on the other two types of waves.

The waves that propagate horizontally along the hole bottom encompass two main components: direct surface waves, specifically Rayleigh waves, and direct body waves. These waves typically reach the hole wall around 0.6 ms after transmission. Over time, the components of the Rayleigh waves gradually convert into body waves as they continue to propagate (as illustrated in Fig. 8.6 at 1.0 ms). Additionally, portions of the direct Rayleigh waves are reflected, subsequently traveling back along the hole bottom. Notably, multiple reflections from the hole wall are observed at various time points (e.g., at 1.6, 2.6, and 3.6 ms). At each reflection point, mutual conversions between Rayleigh waves and body waves occur, resulting in a gradual decay of energy for these multiple reflections. Both the direct waves and the waves stemming from these multiple reflections are captured by the sensors, contributing to the complexity of the test signal.

The waves that propagate downwards into the bedrock encompass P-waves and S-waves. These two types of waves gradually separate due to the significantly higher velocities of P-waves compared to S-waves (as observed in Fig. 8.6 at 0.2 and 0.6 ms, respectively). The separation between the downward-traveling P-waves and S-waves becomes more pronounced after 1 ms (Fig. 8.6, 1 ms). The P-wave reaches the roof of the karst cavity around 1.0 ms. At this point, a portion of the P-wave is reflected as both a PP-wave and PS-wave, which become noticeably distinct after 1.8 ms. The PP- and PS-waves reflected from the cavity roof subsequently reach the sensor positioned at the hole bottom at 2.0 and 2.6 ms, respectively. Consequently, the depth of the top of the cavity can be estimated by analyzing the arrival times of these reflections. The remaining portion of the P-wave penetrates into the cavity. As depicted in Fig. 8.6 at 1.4 ms, the waveform within the clay-filled cavity exhibits a concave shape compared to the convex waveform outside the cavity. This is due to the lower wave velocity in the cavity, which is significantly smaller than that within the surrounding bedrock. Generally, the amplitude of the wave within the cavity is much higher than outside the cavity (refer to Fig. 8.6, 1.8–3.0 ms) owing to the focusing effect within the spherical karst cavity (Dong and Huang 2010). Once the wave reaches the floor of the cavity at 2.0 ms, it splits into two components: one part transmits through the cavity floor into the bedrock but is eliminated by the PML boundary, while the other part reflects from the floor and undergoes further focusing as it propagates upward within the cavity (see Fig. 8.6, 2.2–3.0 ms). Consequently, the amplitude of the wave reflected from the cavity floor is greater than that reflected from the cavity roof. The reflected PP- and PS-waves from the cavity floor arrive at the sensor located at the hole bottom at 4.0 and 4.6 ms, respectively (as observed in Fig. 8.6, 4.0 and 4.6 ms). By analyzing the difference in arrival times between the reflections from the roof and floor of the cavity, the size of the cavity can be estimated. Further details regarding the estimation procedure will be discussed later on.

Figure 8.7 presents the profiles of the vertical and horizontal components recorded by sensors placed in a simulated continuous distribution along the centerline of the hole bottom. In Fig. 8.7a, the horizontal axis represents the horizontal distances between the transmitter and the simulated sensors across an 8 m detection width, while the vertical axis indicates the wave travel times. The two red dotted lines depict the walls of the pile hole, and the positive amplitudes of the modeled signal are depicted in black.

In Fig. 8.7a, it can be observed that the wave transmitted from the center point (Point O) propagates in two directions along the hole bottom. Initially, in the range of OP, the wave comprises both body waves and Rayleigh waves, resulting in an interference pattern. A portion of the wavefield travels through the bedrock along Line P-M–N. By calculating the difference in arrival times at Points M and N ($t_{MN}$ = 0.902 ms) and the corresponding horizontal distance ($d_{MN}$ = 2 m), the velocity is determined as 2217 m/s, indicating an S-wave based on the similarity to the S-wave velocity listed in Table 8.3. Similarly, the wave along Line P-Q is identified as a Rayleigh wave with a velocity of 1713 m/s. The patterns observed in Fig. 8.9b are similar, except for the transmitted wave, which is identified as a P-wave with an estimated velocity of 3918 m/s, instead of an S-wave as shown in Fig. 8.7a. Consequently, the waves propagating along the pile hole bottom can be characterized as the periodic transmission of S- and P-waves, along with the reflections of Rayleigh waves from the hole walls. As the wavefield evolves, there is a gradual decay of wave energy due to the mutual conversion between Rayleigh waves and body waves. Multiple reflections from the hole walls can be observed between the two dotted lines.

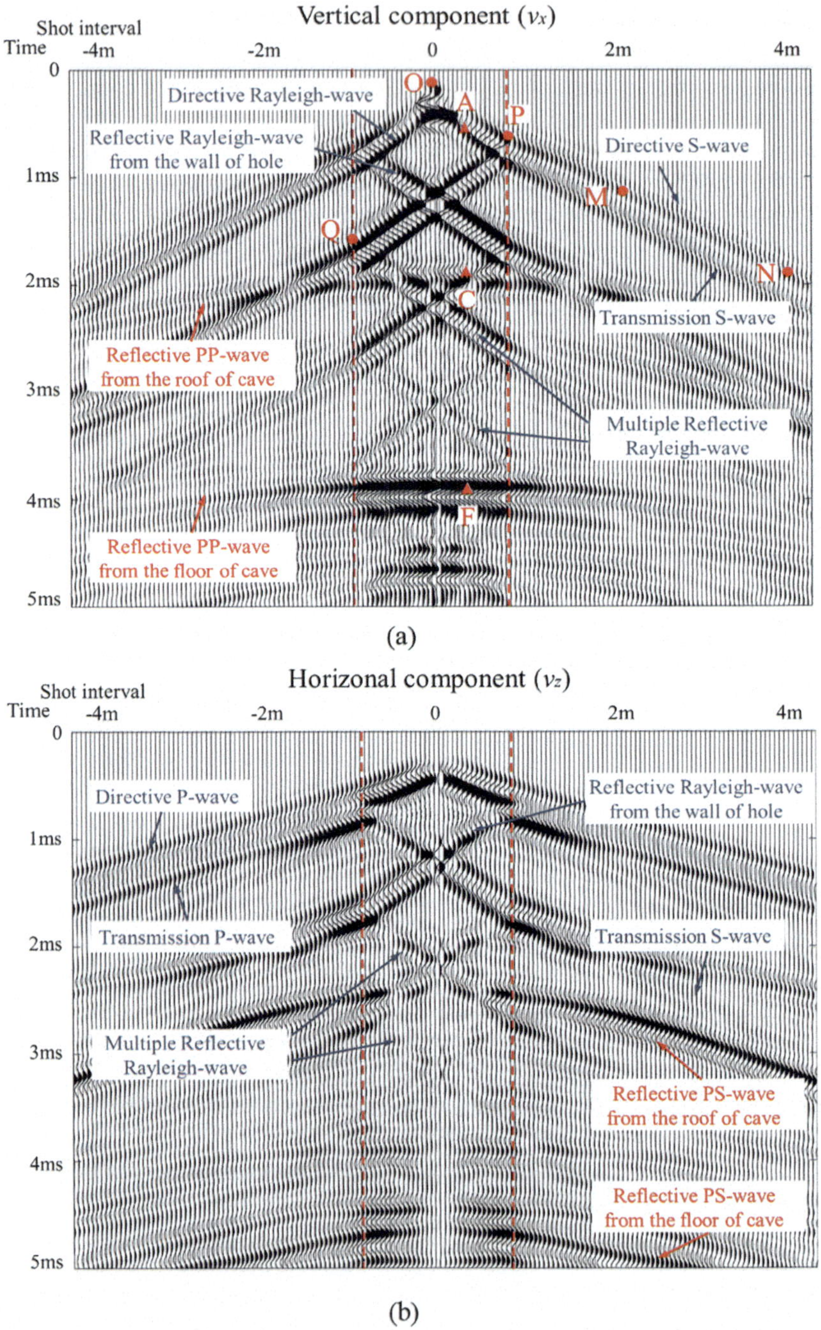


**Fig. 8.7** Wave profile through the centerline of the plane intersecting the hole bottom (Scenario 3): **a** the vertical velocity component ($v_x$) and **b** the horizontal velocity component ($v_z$)

**Table 8.3** Arrival times of the direct and reflected waves

| Arrival time | No. 1 | No. 2 | No. 3 | No. 4 | No. 5 | No. 6 | No. 7 | No. 8 |
|---|---|---|---|---|---|---|---|---|
| $t_{An}$ (ms) | 0.256 | 0.264 | 0.256 | 0.264 | 0.256 | 0.264 | 0.256 | 0.264 |
| $t_{Bn}$ (ms) | 2.112 | 2.084 | 2.042 | 2.014 | 2.020 | 2.050 | 2.090 | 2.118 |

Upon descending into the bedrock, the body waves (S- and P-waves) are only captured by the sensor when they return after being reflected from either the roof or floor of the karst cavity. Figure 8.7a and b illustrate the PP- and PS-waves reflected from the cavity roof and floor, respectively. The reflections from the cavity floor are more prominent than those from the cavity roof due to the aforementioned focusing effects within the spherical cavern. Estimating the depth of the cavity involves utilizing the arrival time of the reflected PP-wave and the known P-wave velocity, or the arrival time of the reflected PS-wave along with the P- and S-wave velocities. By analyzing the difference in arrival times between the reflections from the roof and floor of the cavity, the diameter of the cavity can be estimated to be approximately 2 m. In practical tests, the reflected PP-wave is primarily utilized to calculate the depth and size of the cavity. It is worth noting that during the practical test, the sensor recorded all vertical velocity components and only some of the horizontal velocity components due to its radial angle of approximately 120°. In general, the test signal predominantly consisted of PP-waves reflected from the karst cavity, which greatly interfered with the multiple Rayleigh waves reflected from the hole walls.

(2) **Detection of the Cavity Depth (*H*$_c$)**

Three different numerical scenarios (Scenarios 2, 3, and 4) are considered, corresponding to varying karst cavity roof depths (Hc) of 1, 3, and 5 m. The wave profiles for these scenarios are depicted in Fig. 8.8a (1 m), Fig. 8.7a (3 m), and Fig. 8.8b (5 m). The arrival times of the direct waves are indicated by Point A in Fig. 8.8a, Fig. 8.7a, and Fig. Figure 8.8b, while the reflections from the cavity roof at depths of 1, 3, and 5 m are represented by Point B in Fig. 8.8a, Point C in Fig. 8.7a, and Point D in Fig. 8.8b, respectively. The modeled test signals received by the sensor are displayed in Fig. 8.9, with a spacing of 0.4 m between the transmitter and sensor (d = 0.4 m in Fig. 8.2b). By comparing the signals with and without a cavity, the arrival times at Points A (0.256 ms), B (0.726 ms), C (1.766 ms), and D (2.817 ms) are determined. The depth of the cavity roof is approximately calculated using the P-wave velocity in the bedrock and the difference in arrival times between the reflected and direct waves. The calculated depths for Points B (1 m), C (3 m), and D (5 m) are found to be 0.940, 3.020, and 5.122 m, respectively. It should be noted that the calculation errors may arise from uncertainties in identifying arrival times and the transmitter–receiver interval.

The spectrum of the signal without a karst cavity generally appears smooth (refer to Fig. 8.9b). On the other hand, the test signals with a karst cavity exhibit pronounced oscillations. As the karst cavity becomes shallower, the amplitudes and oscillations in the spectra become stronger, while the reflection energies increase with decreasing

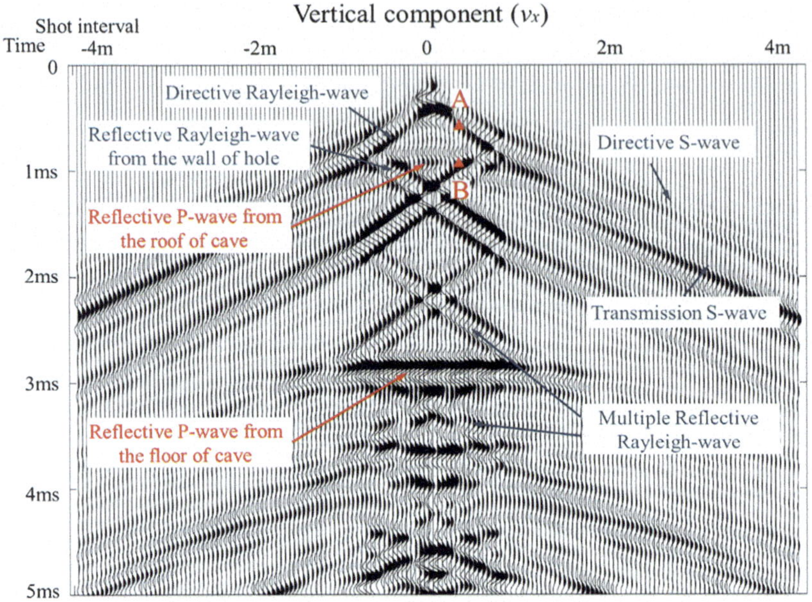


(a)

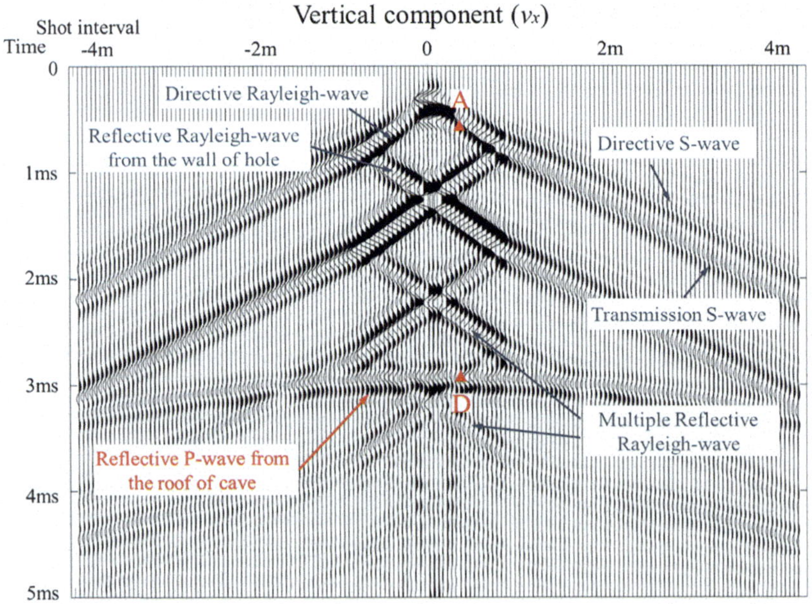


(b)

**Fig. 8.8** Wave profile of the vertical velocity component for different cavity depths: **a** 1 m (Scenario 2) and **b** 5 m (Scenario 4)

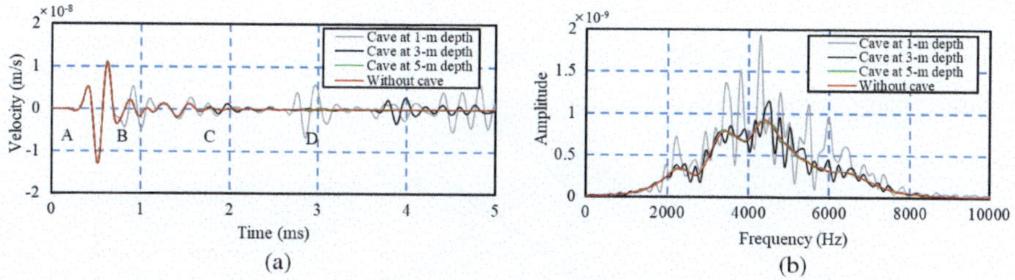


**Fig. 8.9** Modeled signal and the corresponding spectra of different cavity depths

cavity depth. These characteristics can offer valuable insights for qualitative assessments of the bedrock and cavity depth beneath the pile. In the signal shown in Fig. 8.9a, multiple reflections from the cavity floor at a depth of 1 m are observable. In general, the test signals display two prominent peaks in the frequency spectrum, as depicted in Fig. 8.9b. One peak corresponds to the dominant source frequency (approximately 4.4 kHz), while the other peak (approximately 3 kHz) arises from the resonance frequency resulting from multiple reflections from the hole wall. The latter will be examined in further detail later in the chapter.

### (3) **Detection of the Cavity Diameter ($D_c$)**

To investigate the influence of cavity diameter variations (Dc), three different scenarios (Scenarios 5, 3, and 6) with cavity diameters of 1, 2, and 3 m, respectively, are analyzed. The wave profiles for these scenarios are depicted in Fig. 8.10a (1 m), Fig. 8.7a (2 m), and Fig. 8.10b (3 m). The reflections from the cavity roof are denoted as Point C, which remains consistent across the three scenarios. Additionally, Point E in Fig. 8.10a, Point F in Fig. 8.7a, and Point G in Fig. 8.10a represent the reflections from the floor of the cavity with radii of 1, 2, and 3 m, respectively. It can be observed that the size of the cavity affects the arrival times of the reflections from the cavity floor. The modeled test signals recorded by the same sensor are presented in Fig. 8.11a. By utilizing the P-wave velocity of the cavity infill and calculating the differences in arrival times between Point C (1.766 ms) and Points E (2.776 ms), F (3.786 ms), and G (4.666 ms), the cavity sizes can be estimated. The diameters corresponding to Scenarios 5, 3, and 6 are calculated as 1.010, 2.020, and 2.900 m, respectively.

Comparing the reflection from the floor of the 1 m diameter cavity (Point C in Fig. 8.11a and gray line in Fig. 8.11b) with the reflection from the floor of the 2 m diameter cavity (Point F in Fig. 8.11a and black line in Fig. 8.11b), it is evident that the amplitude of the reflection from the floor of the larger cavity is stronger. This can be attributed to the more pronounced focusing effect within the larger cavity. It should be noted that the reflection from the floor of the 3 m diameter cavity is not recorded due to the limited sampling time.

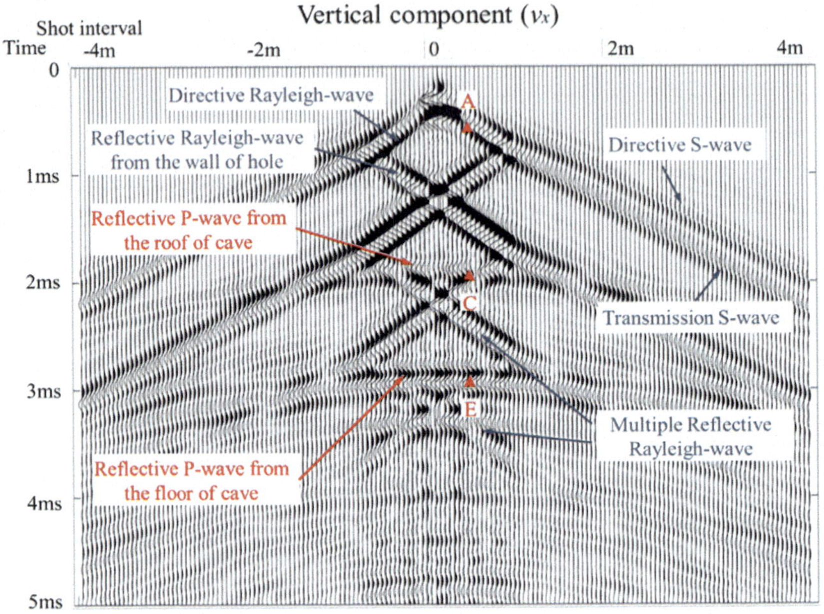


(a)

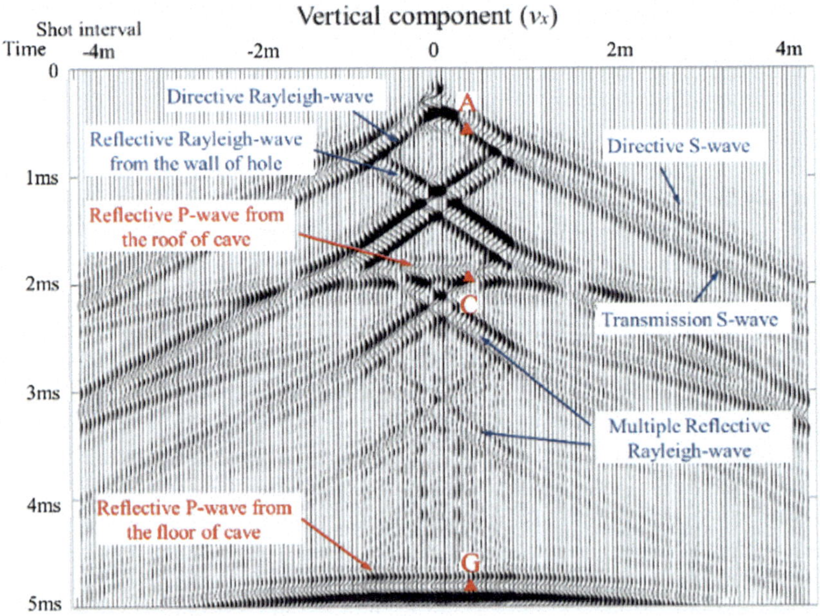


(b)

**Fig. 8.10** Wave profile of the vertical velocity component for different cavity diameters: **a** 1 m (Scenario 5) and **b** 3 m (Scenario 6)

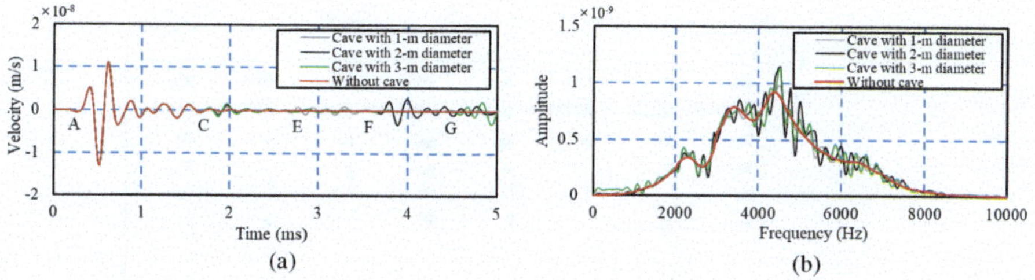


**Fig. 8.11** Modeled signal and the corresponding spectra of different cavity diameters

## (4) **Detection of the Cavity Direction**

In order to demonstrate the location of the karst cavity by rotating the detector, Scenario 7 is considered. In this scenario, a cavity with a diameter of 2 m and a depth of 3 m is positioned between receivers No. 4 and No. 5, as shown in Fig. 8.12. The coordinates of the cavity center are (4, 2, 1), and the directional angle θ is 26.56°.

The test signals received by the eight receivers are combined in the profile depicted in Fig. 8.13. Dotted lines A and B represent the direct wave and the reflection from the roof of the cavity, respectively. Line C represents the reflection from the floor of the cavity, exhibiting a focusing effect. The arrival of the reflected waves at the sensors follows the sequence: No. 4, No. 5, No. 3, No. 6, No. 2, No. 7, No. 8, and No. 1. The convex shape of line B indicates that the cavity is located in the direction between receivers No. 4 and No. 5, as these receivers received the reflected signals earlier than the others.

**Fig. 8.12** Detection of the direction of a karst cavity

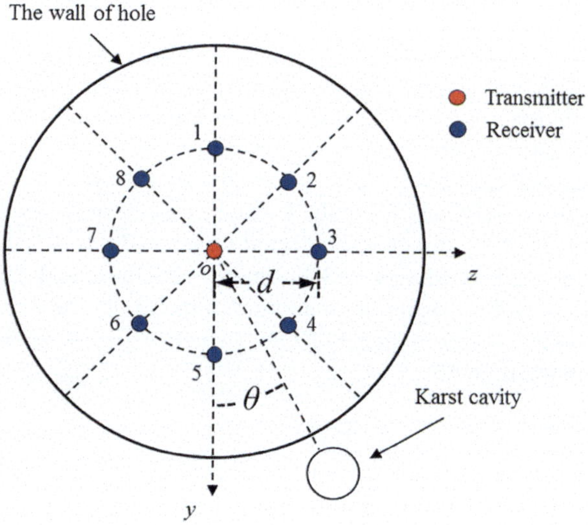

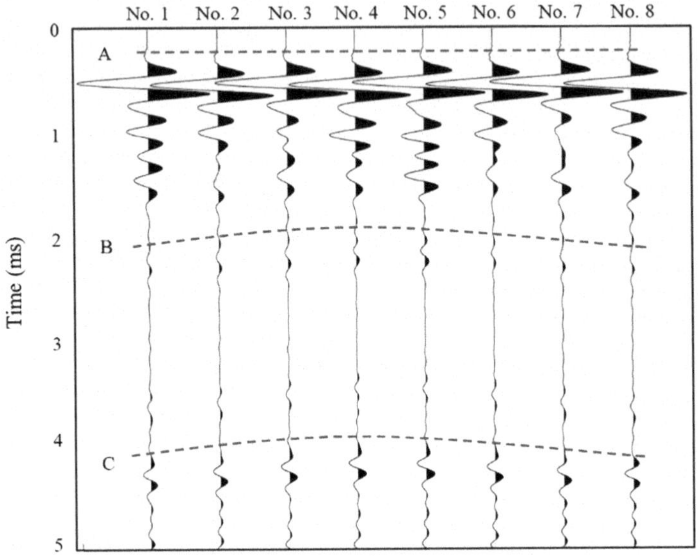


**Fig. 8.13** Wave profiles of eight receivers (Scenario 7)

Assuming that the reflection points on the roof are the same and neglecting the intervals between the transmitter and receivers, the location of the karst cavity roof can be calculated using the arrival times of the reflected and direct waves observed at different receivers based on the following equation: reflected and direct waves observed at different receivers based on the following equation:

$$(x - x_n)^2 + (y - y_n)^2 + (z - z_n)^2 = ((t_{Bn} - t_{An}) \times C_1/2)^2, \tag{8.32}$$

where $(x, y, z)$ denotes the coordinates of the reflector, $(x_n, y_n, z_n)$ $(n = 1, 2, \ldots, 8)$ denotes the coordinates of a receiver (e.g., $(x_1, y_1, z_1)$ is $(0, 0, -0.4)$ in this model), $t_{An}$ and $t_{Bn}$ $(n = 1, 2, \ldots, 8)$ are the arrival times of the direct and reflected waves, respectively, and $C_1$ is the P-wave velocity of the bedrock. Referring to the arrival times of the direct and reflected waves in Table 8.5, a least-squares solution can be obtained, yielding x = 3.473 m, y = 0.836 m, and z = 0.442 m, which correspond to the reflection point on the roof of the cavity. As a result, the estimated directional angle $\theta$ is approximately 27.86°, which is in close proximity to the actual directional angle of 26.56°.

### 8.2.4 Detection Factors Discussion

#### (1) The Optimal Sensor Location

The influence of the shot interval (d) on the reflected surface waves is analyzed in this section using Scenario 3 as an example. Figure 8.14a–e displays the test signals with different shot intervals (d), while Fig. 8.14f shows the wave reflected from the cavity obtained by subtracting the test signal with a cavity from that without a cavity (Fig. 8.14a–e). It can be observed that the reflections from the cavities with different shot intervals are similar, indicating that the shot intervals do not have a significant effect on the reflection from a cavity. Therefore, the influences of the shot interval (d) on the reflected surface waves are discussed.

The locations of Point A in Fig. 8.14a–e correspond to the arrival times of the direct surface waves, while Points B and C represent the multiple reflections of the surface waves from the hole wall. The difference between the arrival time of the direct surface wave and the first reflected surface wave from the hole wall ($t_{AB}$) decreases

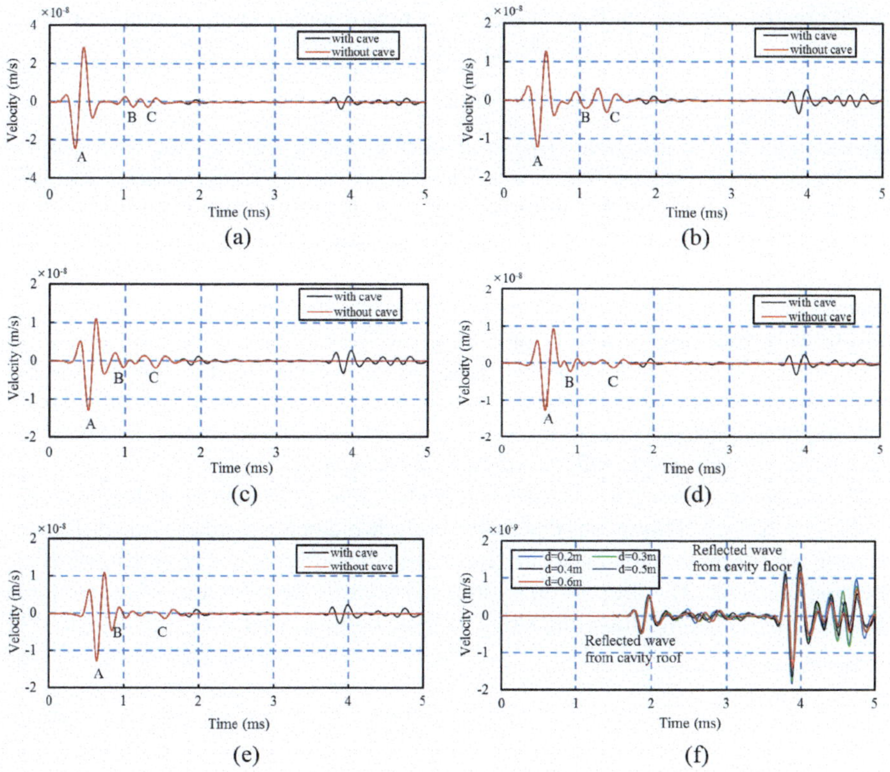


**Fig. 8.14** Modeled signals with different shot intervals (Scenario 3): **a** $d = 0.2$ m; **b** $d = 0.3$ m; **c** $d = 0.4$ m; **d** $d = 0.5$ m; and **e** $d = 0.6$ m. **f** The recorded reflections from the cavity with different shot interval

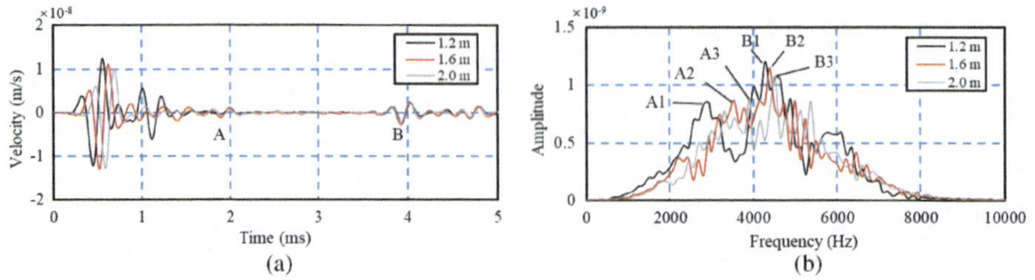


**Fig. 8.15** Modeled signal and the corresponding spectra of different pile hole diameters

with increasing shot interval, whereas the difference between the arrival times of the first two reflected surface waves from the hole wall ($t_{BC}$) increases with the shot interval. When d = 0.4 m (i.e., the shot interval is 0.5 times the radius of the pile), the multiple reflected surface waves exhibit symmetry (i.e., $t_{AB} = t_{BC}$). In such cases, the waves reflected from the cavity can be identified if reflections occur in different time periods. Based on these observations, it may be possible to develop a time–frequency filter to mitigate the interference caused by multiple surface wave reflections from the pile walls in future studies. Therefore, a sensor with a shot interval that is 0.5 times the pile radius can be considered optimal.

(2) **The Diameter of the Pile Hole ($D_p$)**

The diameter of the pile hole (*Dp*) is another factor that affects the arrival times of reflected surface waves. Three scenarios (Scenarios 8, 3, and 9) with pile hole diameters of 1.2, 1.6, and 2.0 m, respectively, are analyzed. The different pile hole diameters result in distinct arrival times of direct waves and multiple surface wave reflections from the hole walls (Points C, D, and E in Fig. 8.15a). As the pile hole diameter decreases, the direct waves arrive earlier, and the period of the multiple reflections from the hole wall becomes shorter.

Figure 8.15b shows the dominant frequencies (Points B1, B2, and B3) of the modeled signals, which depend on the transmitted source frequency. Additional resonant frequency peaks can be observed at Points A1, A2, and A3 in Fig. 8.15b, which are caused by the multiple reflections from the hole wall.

The differences between the two dominant frequencies follows an order of $\Delta f_{A1B1} > \Delta f_{A2B2} > \Delta f_{A3B3}$. This means that as the pile hole diameter decreases, the difference between the resonant frequencies increases, and the difference is proportional to the pile hole diameter. The proportional ratio is similar to the velocities of the multiple surface wave reflections traveling along the hole bottom. Therefore, the velocities of the multiple surface wave reflections can be predicted using Eq. (8.33):

$$v_R = L \times \Delta f, \tag{8.33}$$

where $v_R$ is the velocity of a multiple reflected surface wave, $L$ is the diameter of the pile hole, and $\Delta f$ is the difference between the resonant frequencies. The predicted velocities and their errors are shown in Table 8.4.

**Table 8.4** Estimated velocity of surface wave

| Diameter of pile | $\Delta f$ (Hz) | $v_R$ (m/s) | Estimated $v_R$ (m/s) | Error (%) |
|---|---|---|---|---|
| 1.2 | 1343 | 1713 | 1611 | 5.9 |
| 1.6 | 977 | 1713 | 1563 | 8.8 |
| 2.0 | 854 | 1713 | 1708 | 0.3 |

In the practical test, the velocities of the multiple surface wave reflections along the hole bottom cannot be directly measured by the detector. However, these velocities play a crucial role in predicting the arrival times of the multiple surface wave reflections and identifying the reflections from the cavity. In this study, we propose a novel method for predicting the velocities of these multiple surface wave reflections using the spectrum of the test signal. By estimating the arrival times of the multiple surface wave reflections based on their velocities, we can effectively identify the reflection from the cavity and eliminate the interference caused by reflected surface waves.

(3) **The Transmitted Source Frequency**

In order to investigate the impact of the transmitted source frequency, we studied three scenarios (Scenarios 10, 3, and 11) with varying dominant Ricker wavelet frequencies (2, 3, and 5 kHz). Our analysis revealed that as the transmitted source frequency increases, the energy of the reflection from the cavity floor (Point B in Fig. 8.16a) also increases. This can be attributed to the pronounced influence of the focusing effect on waves with higher frequencies.

The spectra of the modeled signals exhibit two peaks each (peaks A1 and B1, A2 and B2, and A3 and B3 in Fig. 8.16b). Increasing the transmitted source frequencies leads to higher dominant frequencies in the received signals (i.e., fB3 > fB2 > fB1). However, the differences between the two frequency peaks in each spectrum are approximately equal (i.e., $\Delta f_{A1B1}$ (1098 Hz) $\approx \Delta f_{A2B2}$ (977 Hz) $\approx \Delta f_{A3B3}$ (977 Hz)), indicating that the difference between the dominant frequencies is primarily determined by the diameter of the pile hole. The predicted velocities of the multiple surface wave reflections along the hole bottom are 1757 m/s (using $\Delta f_{A1B1}$), 1563 m/s

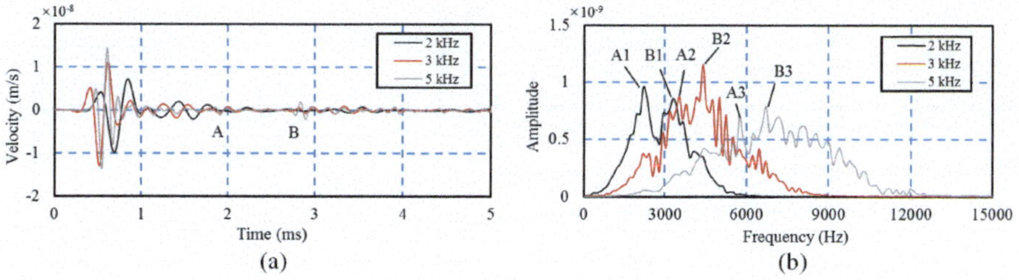


**Fig. 8.16** Modeled signal and the corresponding spectra of different source frequencies

(using $\Delta f_{A2B2}$), and 1563 m/s (using $\Delta f_{A3B3}$). These values closely match the known velocity of the multiple surface wave reflections (1713 m/s), thus validating Eq. 8.33.

(4) **The Cavity Infill**

Three scenarios (Scenario 12, 3, and 13) with different infill materials of clay, water, and air (Table 8.5) are investigated to analyze their effects on the test signal. In Fig. 8.17, it can be observed that the reflection from the cavity roof (Point A in Fig. 8.17a) filled with air is stronger compared to the other scenarios. This difference is attributed to the increased impedance contrast between the infill and bedrock materials, resulting in a higher amount of wave energy being reflected. The reflection from the cavity floor filled with water (Point C) occurs later than that from the cavity floor filled with clay (Point B) due to the different wave velocities of water and clay. The reflection from the floor of the cavity filled with air was not recorded within the limited sampling time, although it may be recorded if the sampling time is extended beyond 15 ms.

Examining the spectrum in Fig. 8.17b in conjunction with the other spectra discussed, the roughness of the curve in the frequency domain is primarily caused by the reflection from the cavity floor. The presence of strong reflected body waves from the cavity roof contributes to the relative roughness of the frequency curve, providing a potential means of identifying the infill material of a karst cavity in practical experiments.

In general, there are no significant differences among the test signals based on the vertical velocity component (P-wave sensor) for different infill types. However, noticeable differences can be observed among the horizontal velocity components (S-wave sensor), as depicted in Fig. 8.18. The reflection from the cavity is much stronger when the cavity is filled with water or air due to the recording of reflected

**Table 8.5** Material parameter of infill of void

| Material parameter | $v_p$ (m/s) | $v_s$ (m/s) | $\rho$ (kg/m$^3$) |
|---|---|---|---|
| Clay | 2000 | 1100 | 1500 |
| Water | 1500 | 0 | 1000 |
| Air | 340 | 0 | 7.9 |

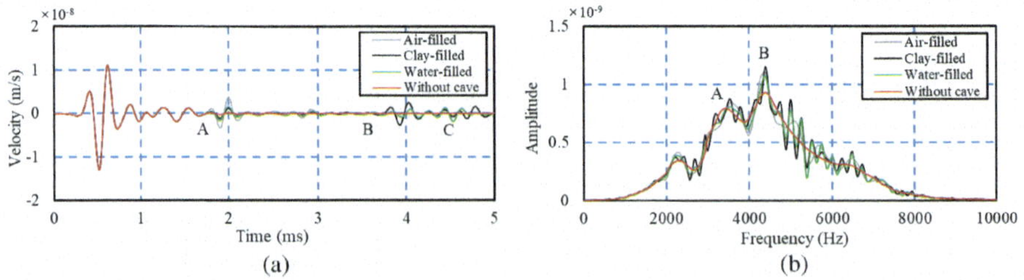


**Fig. 8.17** Modeled signal (the velocity component is in the vertical direction) and the corresponding spectra of different types of infilled cavities

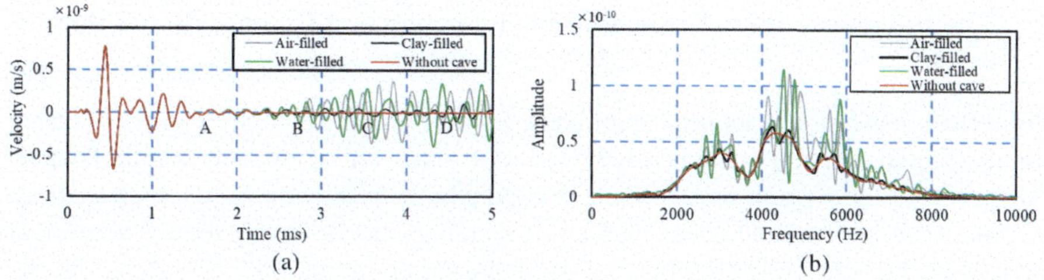


**Fig. 8.18** Modeled signal (the velocity component is in the horizontal direction) and the corresponding spectra of different types of infilled cavities

PS-waves by the S-wave sensor. PS-waves cannot propagate in a fluid, resulting in increased PS-wave energy being reflected when the cavity is filled with water or air. Therefore, it becomes relatively straightforward to identify the infill material of a cavity if the sensor can record the horizontal velocity component. However, current detectors primarily employ P-wave sensors that can only record the vertical component. Hence, the development of S-wave sensors capable of recording the horizontal velocity component is crucial for future research.

This chapter did not consider attenuation in the cavity infill and bedrock due to challenges in calibrating the values of impedance coefficients ($\varepsilon_L$ and $\varepsilon_T$) for different cavity infill and bedrock materials. The reflected waves from the cavity floor can be affected by attenuation in the bedrock (caused by minor fractures) and cavity infill. Determining which material causes the attenuation of the reflected wave is challenging. In Scenario 3, assumed values of $\varepsilon_L$ and $\varepsilon_T$ are 200 and 300 $s^{-1}$ in the bedrock and 1000 and 1500 $s^{-1}$ in the cavity infill, respectively. The modeled signal and corresponding spectra are presented in Fig. 8.19. Notably, the impedance coefficients primarily reduce the amplitude of the reflected wave from the cavity but do not significantly affect the wavefield and cavity identification. Additionally, the simulation of dispersion effects and other factors induced by viscoelastic-porous multiphase media cannot be effectively accomplished using viscoelastic equations alone. Further studies employing additional models and simulation methods that consider viscoelastic-porous multiphase media are needed to investigate wave attenuation and dispersion due to complex infilling materials in karst cavities.

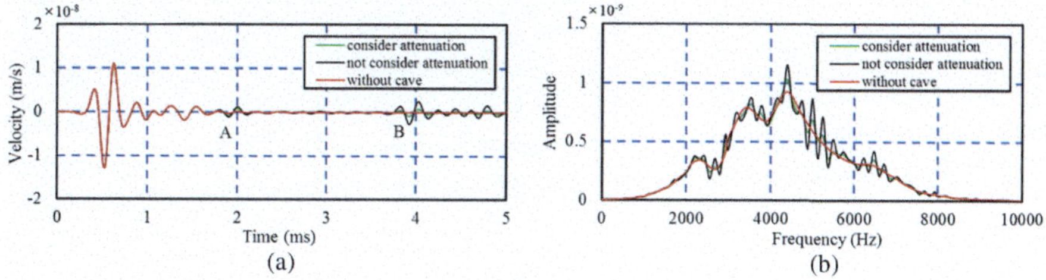


**Fig. 8.19** Modeled signal and the corresponding spectra calculated by FDTD considering the attenuation, or lack of, in bedrock and infill

**(5) The Influences of the Dispersion Phenomenon and Computational Boundaries**

Numerical dispersion is an important consideration in FDTD modeling, particularly when the spatial grid spacing is relatively large compared to the wavelength. Dispersion becomes more pronounced when modeling high-frequency waves, such as those used in sonar detection. Decreasing the grid spacing can effectively reduce numerical dispersion, but this comes at the cost of increased computational resources. In this study, a high-order (tenth-order) FDTD method is employed to address numerical dispersion. The results in Fig. 8.20 demonstrate that the high-order method successfully suppresses dispersion without the need for grid spacing increase, unlike the low-order (fourth-order) FDTD method depicted in the same figure. For this study, a spatial step of 0.05 m and a total grid number of $240 \times 240 \times 240$ are utilized.

The dispersion phenomenon observed in Fig. 8.6 is negligible during wave propagation. This is further validated by the modeling results shown in Fig. 8.20, where the high-frequency wave dispersion is effectively suppressed using the high-order FDTD method without grid spacing augmentation. In this study, the chosen spatial step is 0.05 m, resulting in a total of $240 \times 240 \times 240$ grids. In contrast, a second-order FDTD method would require a spatial step of 0.01 m and a total of $1100 \times 1100 \times 1100$ grids to achieve dispersion elimination. Thus, utilizing a higher-order finite-difference scheme significantly reduces the computational cost.

In this chapter, a simulated space of $8 \times 8$ m is employed to mimic the sonar detection of a karst cavity beneath a pile. It is important to note that the boundaries within the computational domain are not present in practical scenarios. If the boundary

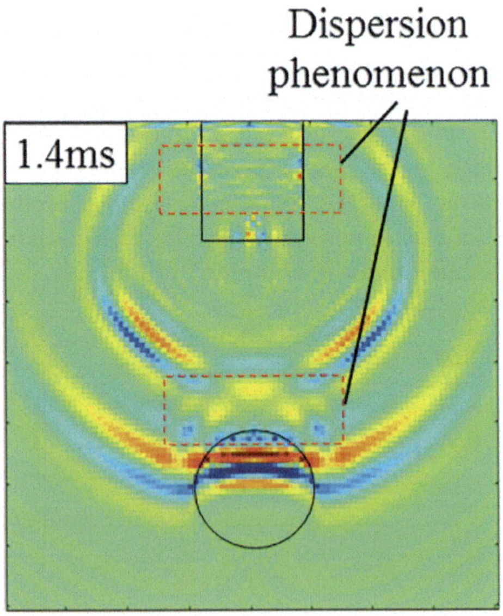


**Fig. 8.20** Snapshots calculated using the low-order FDTD method (Scenario 3)

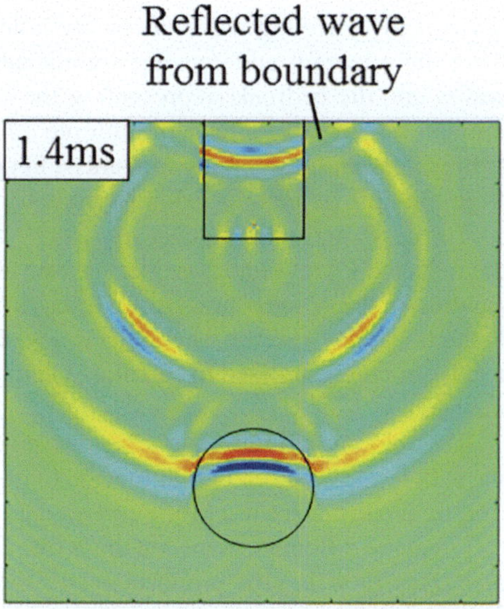


**Fig. 8.21** Snapshot without PML boundaries (Scenario 3)

conditions are not appropriately handled, reflections from these boundaries can significantly impact the modeling outcomes. To address this issue, perfectly matched layer (PML) boundary conditions are applied in conjunction with the surrounding physical domain of the model. The effectiveness of the PML conditions can be observed by comparing the snapshots in Fig. 8.6 with those in Fig. 8.21. As demonstrated in Fig. 8.6, there are no notable reflections from the physical boundaries, and the waves are promptly absorbed upon entering the PML. In reality, reflections from geological interfaces such as joints, faults, or other piles located far from the pile bottom require a longer time to reach the sensor on the pile bottom than the detection period allows. Hence, it is reasonable and effective to model the detection in an infinite space utilizing PML boundary conditions, as this approach prevents unnecessary interference from artificial boundaries. It is important to note, however, that the interference associated with geological interfaces near the pile bottom was not examined in this study and will be a focus of future investigations.

### *8.2.5 Numerical Summary*

This chapter presents a 3D model of a hole-bedrock-cavity system utilizing the Finite-Difference Time-Domain (FDTD) method. The model parameters are calibrated using field experimental data, thereby validating the reliability of the proposed approach. Based on the numerical results, the following conclusions can be drawn:

The waves transmitted from the center of the pile hole bottom can be categorized into three types: upward-traveling waves into the slurry (which can be disregarded), horizontally traveling waves along the hole bottom (generating multiple reflections from the pile wall and causing significant interference), and downward-traveling waves into the bedrock (representing the effective signal for cavern detection). To mitigate the negative effects of multiple reflections, a receiver is strategically placed at the midpoint between the transmission point and the hole wall, where the signals from multiple reflections exhibit symmetry and can be easily identified.

The depth and size of the cavity can be estimated by analyzing the travel time of the reflected P-wave and considering the velocity of the medium. In cases of shallow cavities, multiple reflections from the cavity roof can induce pronounced oscillations in the frequency spectra. The travel times of these reflections, recorded by azimuthal receivers, can be utilized to calculate the direction of the cavity using a least-squares method.

The velocity of multiple surface wave reflections can be predicted by leveraging the difference between two dominant frequencies and the diameter of the pile hole. The frequency spectrum of the test signal typically exhibits two dominant frequencies, and the difference between these frequencies correlates with the pile hole diameter. This ratio is also proportional to the velocity of the multiple surface wave reflections. By predicting the velocity, it becomes feasible to estimate the arrival times of these reflections and identify the reflected P-waves.

The presence of a cavity induces a wave focusing effect when it is completely filled with a low-velocity medium. The energy of the reflected wave from the floor of the cavity is theoretically stronger than that from the roof. The amplitude of the reflected wave increases with a greater impedance contrast between the bedrock and the cavity infill material. Notably, PS-waves reflected from the cavity roof exhibit greater significance than PP-wave reflections due to the inability of PS-waves to propagate through gases or liquids.

The sonar detection of a karst cavity beneath a bored pile can be effectively simulated using the 3D Staggered-Grid Finite-Difference (SGFD) method. The adoption of perfectly matched layer (PML) conditions ensures the absorption of reflections from artificial boundaries. Additionally, numerical dispersion phenomena can be minimized by employing a high-order FDTD method, thereby reducing computational costs. The proposed numerical approach, combining PML conditions and a high-order finite-difference scheme, demonstrates satisfactory reliability in simulating sonar detection of karst cavities under bored cast-in-situ piles.

## 8.3 Signal Qualitative Analysis

### *8.3.1 Qualitative Analysis Method*

Based on the analysis of the numerical simulation results, it is known that the main components of the sonar detection signal for the cavity beneath the bottom of the pile are the reflected waves from the cavity and the multiple reflections of surface waves from the bottom of the pile. In certain specific conditions, such as when the mud density at the bottom of the hole is high, the sediment content is significant, and the pile hole diameter is large, the reflections from the pile wall may not be apparent. If we assume that the interference from the pile wall reflections can be neglected, a rapid qualitative assessment of the cavity development at the bottom of the pile can be made on-site. Since the distance between the sonar transmitter and receiver is very small, it can be considered as a negligible offset distance. The one-dimensional wave theory can be employed for a quick qualitative analysis of the reflected waveforms.

Using the one-dimensional particle longitudinal vibration equation:

$$\frac{\partial^2 u}{\partial t^2} = c^2 \frac{\partial^2 u}{\partial x^2}, \tag{8.34}$$

where u represents the particle's vibration position, t represents time, and x represents the direction of vibration. The solution is given by the D'alembert solution:

$$u(x, t) = f(x - ct) + g(x + ct), \tag{8.35}$$

where $C$ represents the propagation velocity of the wave in the $x$ direction. This equation demonstrates that the vibration of the particle is a superposition of two counter-propagating waves. If we consider $f(x\text{-}ct)$ as the initial downward wave, then $g(x + ct)$ represents the reflected wave from the interface between different media at the bottom of the pile. These waves maintain their original waveform and propagate and superimpose with a velocity of $c$. When the sonar stress wave propagates to the interface of different elastic media, it causes wave emission and refraction. The reflection follows two conditions at the interface: (1) the stress must be equal on both sides of the interface, and (2) the particle velocity must be continuous at the interface, Thus, we have:

$$p + p_R = p_T, \tag{8.36}$$

$$v + v_R = v_T. \tag{8.37}$$

In the above equation: $P$, $PR$, and $PT$ represent the incident, reflected, and transmitted sound pressures, respectively, while $v$, $v_R$, and $v_T$ represent the incident, reflected, and transmitted wave velocities. It is known that $V = P/Z$, where $Z = \rho C$, and $Z$ represents the acoustic impedance of the interface between different media, which is the product of the density ρ and the wave velocity C of the respective medium. Substituting this equation into Eq. 8.37, we have:

$$\frac{P}{Z_1} - \frac{P_R}{Z_1} = \frac{P_T}{Z_2}. \tag{8.38}$$

By combining Eq. 8.36 and 8.38, we obtain the reflection coefficient:

$$R = \frac{P_R}{P} = \frac{Z_2 - Z_1}{Z_1 + Z_2}. \tag{8.39}$$

The obtained reflection coefficient $R$ can be used to analyze the impedance differences between different layers beneath the pile bottom, thereby quickly determining the integrity of the pile bottom, the presence of cavities, the variation in soil stiffness, and fractured rock layers, as shown in Table 8.6. The larger the absolute value of $R$, the greater the impedance difference between the upper and lower media.

When the bedrock beneath the pile bottom is intact, the time-domain curve of the reflected wave received by the pile-bottom sonar transducer exhibits exponential decay with uniform amplitude attenuation, without any secondary reflections. In this ideal condition, $R = 0$, and the signal energy is strong, as shown in Fig. 8.22a.

**Table 8.6** Pile hole sonar detection analysis

| Signal characteristic | Reflection coefficient $R > 0$ | Reflection coefficient $R < 0$ | Reflection coefficient $R = 0$ while large amplitude | Reflection coefficient $R = 0$ while large amplitude |
|---|---|---|---|---|
| Detection results | Soft layer overlying bedrock | Voids beneath pile hole | Intact bedrock | Broken bedrock |

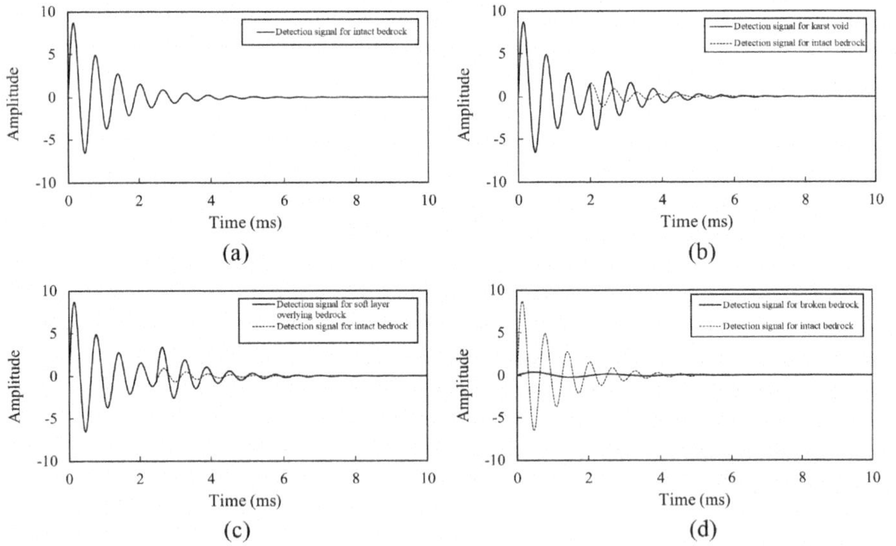


**Fig. 8.22** Ideal reflection signals in pile-bottom cavity detection (ignoring pile wall reflections): **a** intact pile bottom; **b** pile bottom with cavities; **c** pile bottom with softer soil above and harder rock below; **d** extremely fragmented pile-bottom bedrock

When there are cavities developed beneath the pile bottom, the impedance of the filling material in the cavities is smaller than that of the bedrock above the cavities. Therefore, the reflection coefficient $R < 0$, and the phase of the reflected wave is opposite to that of the incident wave beneath the pile bottom. In the time-domain curve, this is manifested as the appearance of secondary reflections and the superposition of an out-of-phase waveform at the location of the cavities, as shown in Fig. 8.22b.

In the case of encountering a harder rock layer beneath the pile bottom, resulting in a condition of softer soil above and harder rock below, the density and impedance of the lower media are higher. Consequently, the reflection coefficient $R > 0$, and the phase of the reflected wave at the interface is opposite to that of the incident wave beneath the pile bottom. In the time-domain curve, this is reflected as the superposition of an in-phase waveform at the interface, as shown in Fig. 8.22c.

When the rock layers beneath the pile bottom are extremely fragmented, there is no clear and distinct reflection surface in the bedrock. The received reflected signal is very weak compared to the signal when the bedrock is intact, resulting in $R = 0$, and the signal energy is weak, as shown in Fig. 8.22d.

### 8.3.2 Field Application

#### (1) A Field Case in Nanning, Guangxi

A commercial project in Nanning, Guangxi, is located in the Wuxiang New District. The proposed main building has a height of 48 floors and is 150 m tall. It includes two underground levels, and the northern and eastern sides facing the street have two levels of commercial shops. The building structure is planned to use shear walls. The site is characterized by residual karst landform, with local variations in terrain. The geological layers at the site, from top to bottom, are described as shown in Table 8.7.

Based on the investigation and drilling data, the underlying bedrock at the site consists of soluble carbonate limestone. The karst features mainly include caves, grooves, and small erosional holes. The depth and scale of karst features vary within the exploration range, with burial depths ranging from 19.10 to 32.50 m and cave heights ranging from 0.90 to 1.70 m. Among the 79 exploration holes drilled into the limestone, four of them revealed caves of different depths and sizes, resulting in a cave encounter rate of 5.33%. The total length of limestone drilled was 622.6 m, with a cumulative cave height of 5.30 m, and the linear karst rate was 0.85%. The development of karst features in the area is relatively weak, and there are localized erosion phenomena in the fractured and intact limestone layers. Groundwater is present within the exploration depth range, indicating the potential for further karst development under the influence of groundwater.

Considering the above karst characteristics, the proposed site falls under the category of weak karst development, with relatively weak karst development patterns. The overlying soil layer is thin in certain areas, and the presence of caves has a

**Table 8.7** Characteristics of the soil and rock layers at a construction site in Nanning

| Formation period | Soil layer | Profile image | Soil layer characteristics |
|---|---|---|---|
| $Q_4^{ml}$ | Loose fill layer | 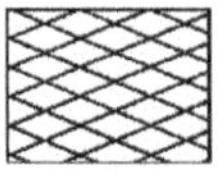 | Predominantly cohesive soil with a small amount of crushed stone, with localized large boulders. Average thickness is 5.37 m |
| $Q_p^{al}$ | Gravelly silt clay layer | 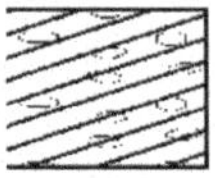 | Predominantly silty clay with gravel composition, mainly consisting of limestone. Average thickness is 10.87 m |
| $C_1y$ | Fractured limestone layer | 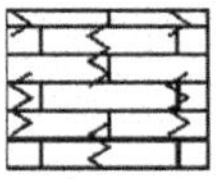 | Developed fractures, with localized dissolution phenomena, moderately weathered. Average thickness is 3.09 m |
| $C_1y$ | Intact limestone layer | 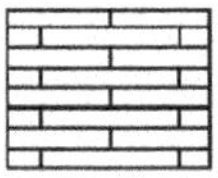 | Slightly weathered, with localized dissolution phenomena, relatively intact rock mass. Average thickness is 7.93 m |

significant impact on the foundation engineering of high-rise buildings. Therefore, pile foundations are recommended for the site.

Figure 8.23a shows the sonar detection signal of a karst cavity at the bottom of pile ZK461 in floors D1-1 to D1-3 of the project. From the bold waveform in Fig. 8.23a, it can be observed that there is a secondary reflected signal with a negative phase at approximately 2 m from the pile bottom, with azimuth angles ranging from 338 to 106°. This indicates $R < 0$, confirming the presence of a cavity to the north of the pile. Further drilling was conducted in that direction, and mud leakage was observed in the borehole, validating the sonar detection result. On-site, the cavity was gradually penetrated using short-range high-frequency impacts. When the cavity roof was breached, mud was immediately replenished into the borehole, the drill was lifted to the borehole mouth, and gravel, clay blocks, and cement were inserted into the borehole to fill the cavity. After stabilizing the water head inside the borehole, a protective wall was formed to contain the cavity, and construction continued.

Figure 8.23b shows the sonar detection signal of a karst cavity at the bottom of pile ZK477 in floors D1-1 to D1-3 of the project. From Fig. 8.23b, it can be observed that the reflection wave signal is stronger on the eastern side of the pile, with no significant secondary reflection observed, indicating intact bedrock. However, the reflection signal on the western side of the pile is very weak, and even with increased amplification, it is difficult to observe clear reflection signals, indicating $R = 0$ and weak signal energy. This suggests that the bedrock on the western side is highly fractured, resulting in significant energy loss of the sonar stress wave. Multiple drilling operations were conducted in various directions beneath the pile bottom, and the cores from the western side exhibited fragmented rock formations, with a rock quality designation (RQD) ranging from 15 to 30. According to Table 6.7, the quality

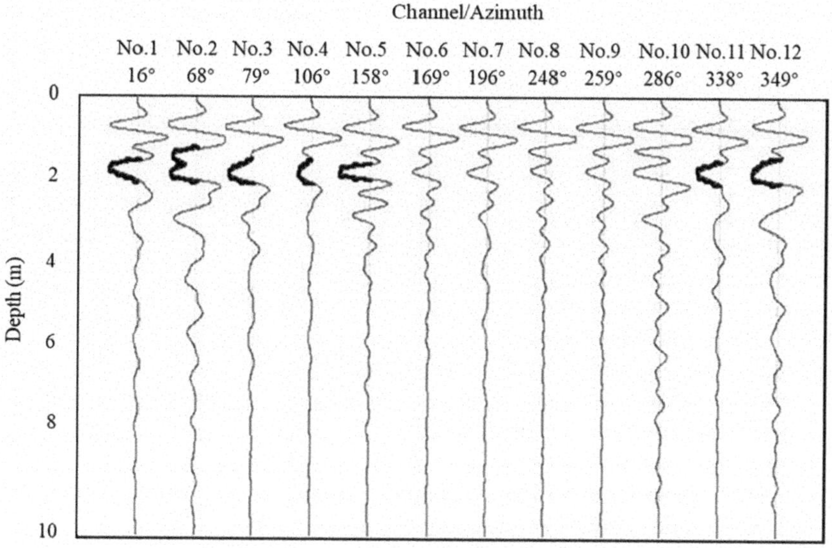


(a)

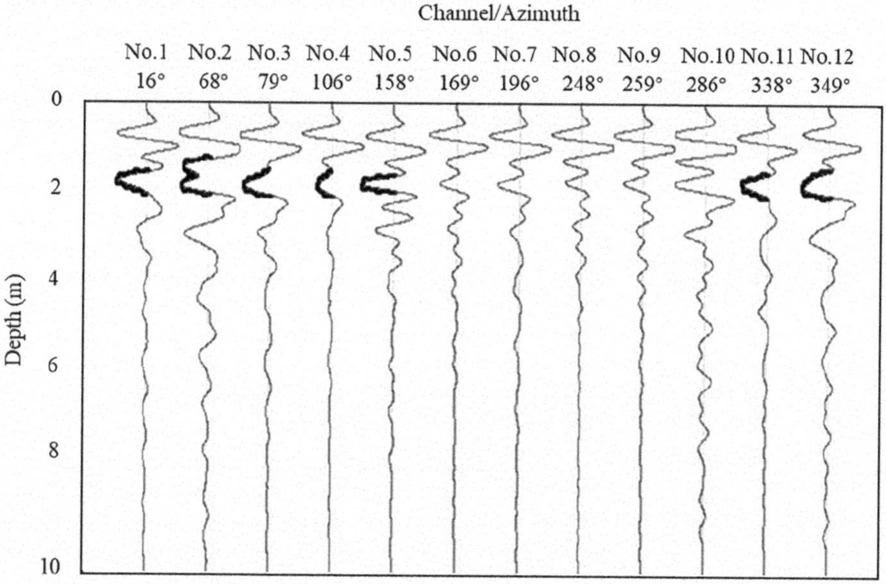


(b)

**Fig. 8.23** Sonar detection signals of ZK461 and ZK477. **a** ZK461; **b** ZK477 of the bedrock beneath the pile bottom is rated as extremely poor, which confirms the results of the sonar detection. On-site, continued drilling was performed until reaching the intact bedrock surface as a remedial measure.

(2) **A Field Case in Anshun, Guizhou**

The project in Guizhou involves the construction of 1# to 5# logistics warehouses and office buildings, ranging from 5 to 12 floors, using a frame-shear structure with column foundations. The site is located in Pingba Town, Anshun City, Guizhou Province. The site is situated in the Yangtze Proto-platform's northern Guizhou tectonic deformation zone, characterized by complex structural deformation. Faults and folds are not well developed in the area, and the geological structure is relatively simple. Based on regional geological information, site surface surveys, and analysis of drilling data, no faults were found to pass through the site. The underlying bedrock is a monoclinal structure with a dip angle of 160° < 20°. The surface of the site consists of brown-yellow fill soil and slope-deposited clay, while the underlying bedrock is thick-bedded dolomite from the middle part of the Lower Triassic Anshun Formation. The geological strata from top to bottom are described in Table 8.8.

Among the 47 drill holes conducted during construction, 11 encountered cavities, resulting in a cavity occurrence rate of 23.4%. The height difference of the bedrock surface ranges from 0.8 to 3.5 m, indicating a moderate development of karst features. The revealed depths of the cavities on the site range from 3.9 to 9.5 m, with vertical heights generally ranging from 0.5 to 2.5 m and the largest cavity reaching 4.7 m. Considering the large building loads and the deep groundwater level, the foundation design specifies a depth of at least 0.5 m below the cavity floor.

Figure 8.24a depicts the detection signal of a cavity at the bottom of pile ZK3-2 in Warehouse 3. The highlighted waveform in the figure clearly shows a negative phase secondary reflection signal of the cavity, indicating R < 0. Further drilling beneath the pile bottom confirmed the presence of the cavity, as significant mud loss was observed in the hole. This observation validates the results obtained from the

**Table 8.8** Characteristics of the soil and rock layers at a construction site in Anshun

| Formation period | Soil layer | Profile image | Soil layer characteristics |
|---|---|---|---|
| $Q_4^{ml}$ | Loose Fill Layer | 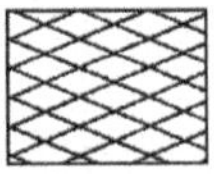 | Predominantly cohesive soil with a small amount of gravel and occasional large rocks, with an average thickness of 5.37 m |
| $Q_p^{al}$ | Red clay |  | Predominantly silty clay with limestone as the main component of the gravel, with an average thickness of 10.87 m |
| $T_1a$ | Dolomite layer | 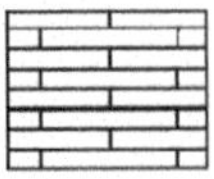 | Slightly weathered, with localized dissolution phenomena and relatively intact rock formation, with an average thickness of 7.93 m |

sonar detection, confirming the existence of a cavity to the east of the pile bottom. After breaking through the cavity ceiling, it was found that the cavity had a height of approximately 1 m. The cavity was initially filled with crushed stone, sand mixture, and cement slurry, followed by compacting the filling with small impact forces to form a mudstone outer wall. Once the wall reached sufficient strength, the drilling continued to penetrate through the cavity.

Figure 8.24b shows the detection signal of a cavity at the bottom of pile ZK3-12 in Warehouse 3. As seen in Fig. 8.24b, the reflected wave signal beneath the pile bottom exhibits uniform attenuation without any apparent secondary reflection, indicating R = 0. The signal energy is strong, suggesting the bedrock beneath the pile bottom is intact without any adverse geological phenomena such as cavities or fractures. No mud loss was observed during drilling, and the rock cores obtained from the drilling appeared as short columns. The rock quality index RQD ranged from 75 to 90, indicating a relatively good quality of the pile-bottom bedrock according to Table 6.7. This finding confirms the results obtained from the sonar detection.

## 8.4 Signal Analysis Based on Generalized S-Transform

Although qualitative analysis of the sonar detection signals of pile-bottom cavities can provide initial interpretation on-site, this process relies on the experience of the analyst and can only be applied under specific conditions where the influence of the pile wall surface waves can be neglected. Therefore, it is necessary to introduce a better method for analyzing the sonar detection signals of pile-bottom cavities to extract the arrival time of the reflected waves from the cavities.

The generalized S-transform (GST) (Pinnegar and Mansinha 2004) is a time–frequency method for analyzing geophysical exploration data. It provides independent frequency resolution while maintaining a close connection to the Fourier transform. The time–frequency distribution of weak signals can be accurately described in the time–frequency spectrum using the GST. In this section, we utilize this characteristic of the GST to identify the arrival time of the reflected waves from pile-bottom cavities in the detection signals. This enables the calculation of the depth of the pile-bottom cavity or other geological bodies. Numerical simulation signals and field detection signals will be used to demonstrate the effectiveness of the GST in analyzing sonar detection signals of pile-bottom cavities.

### *8.4.1 Signal Analysis Method Based on S-Transform (GST)*

The generalized S-transform (GST) of any detection signal $x(t)$ is represented by Eqs. 8.40–8.42, as follows:

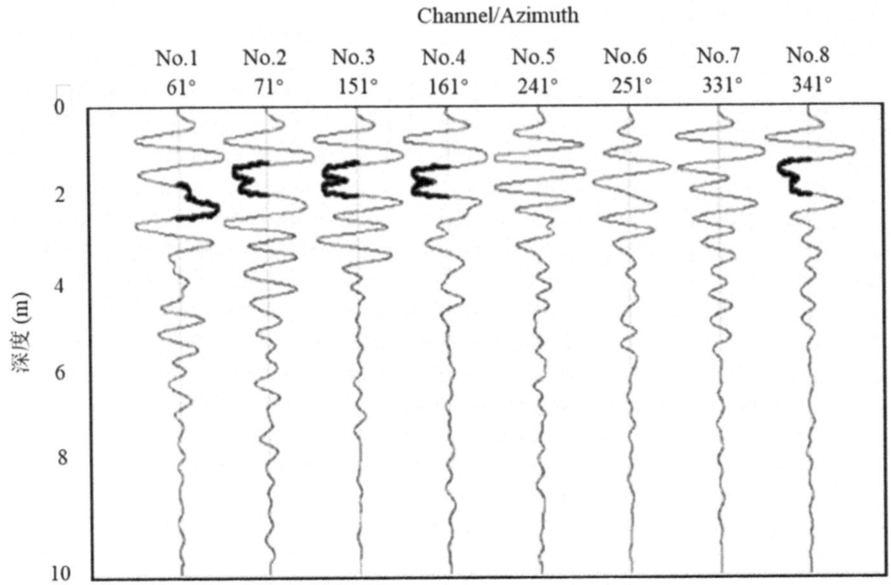


(a)

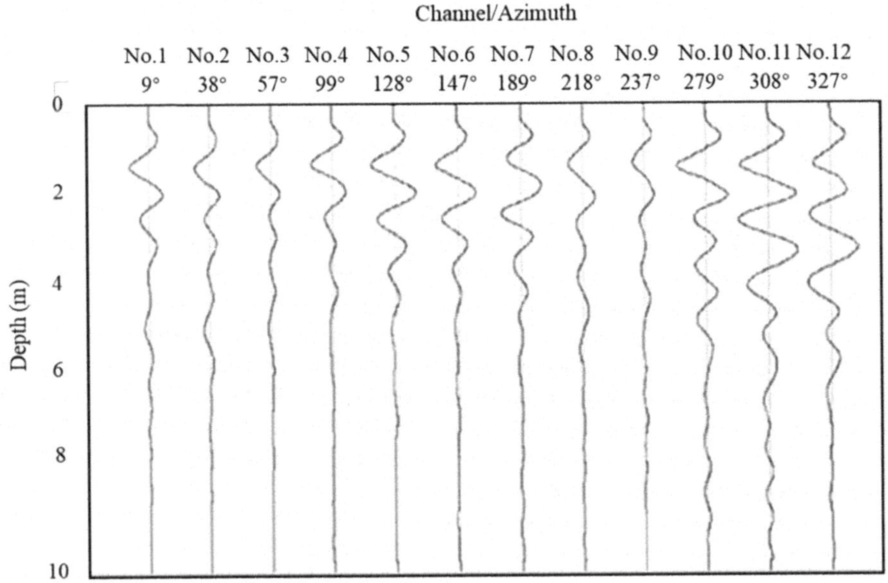


(b)

**Fig. 8.24** Sonar detection signals of ZK461 and ZK477 **a** 3–2, **b** 3–12

$$S(\tau,f) = \int_{-\infty}^{+\infty} x(t)\omega(\tau - t,f)e^{-i2\pi ft}dt, \quad (8.40)$$

$$\omega(\tau,f) = \frac{|f|^p}{k\sqrt{2\pi}}\exp\left[-\frac{f^{2p}\tau^2}{2k^2}\right], \quad (8.41)$$

$$\sigma(f) = \frac{k}{|f|^p}, \quad (8.42)$$

where $\tau$ and t represent time, and $f$ represents frequency. $\omega$ is the Gaussian window function used in the generalized S-transform. $\sigma(f)$ is the scale factor of the Gaussian window function, $p$ is the correction factor, and $k$ is the correction parameter.

To achieve better time–frequency resolution, the correction factor and parameters are introduced into the scale factor of the Gaussian window function. The width of the window function increases with increasing k and decreasing p. Overall, the correction factor $p$ has a greater impact on the shape (width and height) of the window function compared to the correction parameter $k$. Generally, a wider time window can be used for low-frequency signals to obtain higher frequency resolution, while a narrower time window can be used for higher time resolution at high frequencies.

Based on the Fourier transform, convolution theorem, and formulas mentioned above, the generalized S-transform can be expressed as:

$$GST(\tau,f) = \int_{-\infty}^{+\infty} X(\alpha+f)e^{-\frac{2\pi^2\alpha^2k^2}{f^{2p}}}e^{-i2\pi\tau t}d\alpha, \quad (8.43)$$

where $X(\alpha + f)$ is obtained by utilizing the Fourier transform and shifting on $x(t)$. We can derive the discrete generalized S-transform from Eq. 8.44 as follows:

$$GST[j,n] = \sum_{m=0}^{N-1} X\left[\frac{n+m}{NT}\right]e^{-\frac{2n^2k^2m^2}{n^{2p}}}e^{\frac{i2\pi mj}{N}}, \quad (8.44)$$

where $f = n/NT$, $\tau = jT$, $T$ and $f$ are the sampling interval and sampling frequency, respectively. The time–frequency distribution of the detection signal can be calculated using Eq. 8.44.

When there are cavities or rock fractures beneath the pile base, the reflection waves from the cavity ceilings, cavity floors, and fractures will alter the waveform of the detection signal. The changes in the detection signal in the time domain may be subtle, but they will manifest as abnormal peaks in the frequency or amplitude in the time–frequency spectrum. By identifying the arrival times corresponding to these abnormal peaks, we can quantitatively analyze the depth, size, and other information of the cavities or fractures. After identifying the reflection points of the cavity ceilings and floors in the time–frequency spectrum, we can extract their time coordinates $t_A$ and $t_B$ to calculate the depth, size, and other information of the cavities or fractures.

In practice, we place the sonar transmitter transducer at the center of the pile base and symmetrically arrange four signal receivers along the centerline. The distance between the receivers and the transmitter is denoted as $\Delta l$. For the two symmetric receivers, the extracted travel times of the reflection waves in the detection signal can be denoted as $t_{IA}$, $t_{IB}$, $t_{2A}$, and $t_{2B}$, respectively. Assuming the longitudinal wave velocity in the bedrock is $v_{pI}$ and the wave velocity in the cavity filling material is $v_{p2}$, and considering x as the depth of the reflection point on the cavity ceiling from the pile base and $z$ as the horizontal distance, and $r$ as the radius of the cavity, the following equations can be obtained through geometric relationships:

$$\left(\frac{V_{p_1}t_{1A}}{2}\right)^2 = x^2 + \left(z - \frac{\Delta l}{2}\right)^2, \tag{8.45}$$

$$\left(\frac{V_{p_1}t_{2A}}{2}\right)^2 = x^2 + \left(z + \frac{\Delta l}{2}\right)^2. \tag{8.46}$$

Then we have:

$$z = \frac{(V_{p_1}t_{2A})^2 - (V_{p_1}t_{1A})^2}{8\Delta l}, \tag{8.47}$$

$$x = \sqrt{\left(\frac{V_{p_1}t_{1A}}{2}\right)^2 - \left(z - \frac{\Delta l}{2}\right)^2}, \tag{8.48}$$

$$r = \frac{Vp_2(t_{1A} - t_{1B})}{4}. \tag{8.49}$$

Based on this, we can first extract the arrival time of the weak reflection from the cavity at the pile base using the generalized S-transform. Then, by using Eqs. 8.47–8.49, we can quantitatively calculate the size and position of the cavity at the pile base.

### 8.4.2 Numerical Tests

In this section, we conducted two-dimensional numerical simulations to generate a large number of detection signals in order to test the effectiveness of the generalized S-transform in identifying reflected waves from cavities. The numerical simulations used the same FDTD program with material parameters consistent with the three-dimensional case. The only difference was that in order to increase the difficulty of identifying the reflected waves, the attenuation of the mud in the pile hole was set to be lower. As a result, the multiple reflections from the pile wall became more pronounced, leading to increased background noise and posing a challenge in identifying the weak reflected waves at the pile base. In the numerical simulations, the diameter of the pile hole was 1.6 m, and the distance between the transmitting and receiving transducers was 0.4 m. Other material parameters are shown in Table 8.9.

### (1) **Typical Characteristic Signals**

Figure 8.25a shows the detection signal when there are no cavities or other reflections at the pile base. Figure 8.25b shows the corresponding time–frequency spectrum obtained using the generalized S-transform. As the energy of the wall reflection waves gradually decays, frequency peak points appear at regular intervals and their magnitudes decrease. The signal energy gradually attenuates without any significant frequency or amplitude increase, indicating that there are no apparent reflections at the pile base and thus the pile base rock is intact.

**Table 8.9** Material parameters for numerical simulation

| Materials | $v_P$ (m/s) | $v_s$ (m/s) | $\rho$ (kg/m$^3$) | $\varepsilon_L$ | $\varepsilon_T$ |
|---|---|---|---|---|---|
| Bedrock | 4000 | 2300 | 2300 | 0 | 0 |
| Slurry in pile hole | 2000 | 0 | 1500 | 1500 | 2500 |
| Clay in voids | 2000 | 1100 | 1500 | 1000 | 1500 |

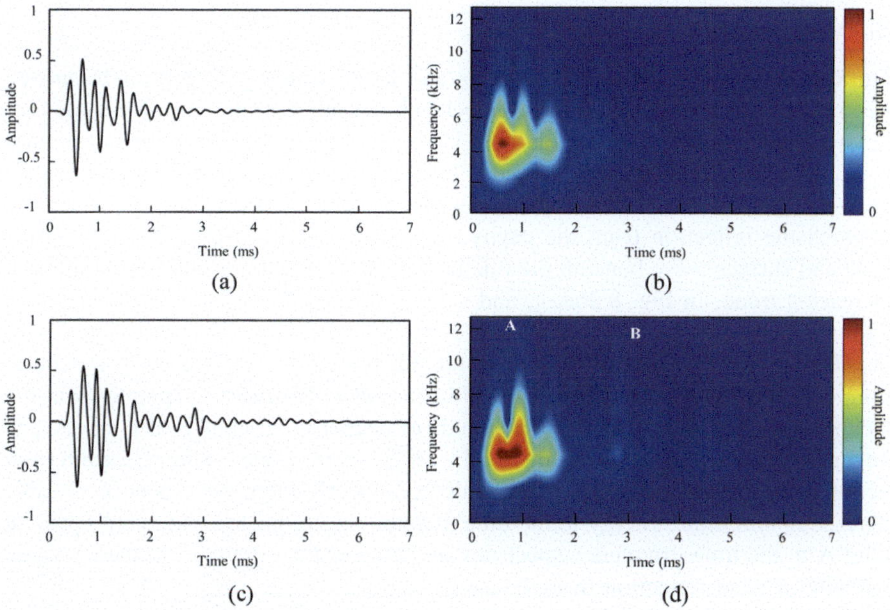


**Fig. 8.25** Typical characteristic signal analysis: **a** Original detection signal without cavity beneath the pile bottom; **b** corresponding time–frequency spectrum using generalized S-transform for the pile bottom without cavity; **c** original detection signal with a cavity at a depth of 3 m beneath the pile bottom; **d** time–frequency spectrum using generalized S-transform for the pile bottom with a cavity at a depth of 3 m

Figure 8.23c and d represents the detection signal with a cavity at the pile base and its corresponding time–frequency spectrum. The cavity's top is located at a depth of 3 m, and its radius is 1 m. From the original signal, it can be observed that the reflected waves from the cavity's top are masked by the direct waves and multiple surface waves reflected from the pile wall, making it difficult to identify them directly from the original signal. However, the reflection from the cavity's bottom can be identified at around 4 ms in the original signal.

In the time–frequency spectrum of the generalized transform, the regularly spaced frequency peak points are interrupted at point A in Fig. 8.25d. Point A appears earlier than the expected regularly spaced peak points, and the frequency does not show a significant decrease. Additionally, there is a gap at the expected regularly spaced peak points. This indicates that point A corresponds to the reflection from the cavity's top. Point B corresponds to the reflection from the cavity's bottom, with a slightly lower frequency peak compared to point A.

The regular amplitude energy decay and regularly spaced frequency peak points, which are not clearly visible in the time domain, can be well identified in the time–frequency spectrum of the generalized S-transform. Therefore, this method is important for identifying cavities at the pile base. Next, we will test the effectiveness of this method by changing the depth, size, horizontal position of the cavities, and their combination with fissures.

(2) **Varying Depths of Karst Voids**

Figure 8.26 shows the detection signals and their corresponding time–frequency spectra for different depths of cavities beneath the pile bottom, with the same size and horizontal position of the cavities. When the cavity is shallow, the reflection from the cavity's top plate arrives earlier. In the time domain, it is difficult to identify due to the interference of multiple reflection surface waves. In Fig. 8.26a and c, although the reflection from the cavity's top plate is superimposed on the direct wave and surface wave beneath the pile bottom, the overall amplitude still shows a decreasing trend. In Fig. 8.26e, g, and i, the reflection from the cavity's top plate gradually separates from the interference of the surface wave, but this depends on the attenuation coefficient of the mud in the pile hole. If the surface wave attenuates more slowly in the mud, deeper reflections are also difficult to separate from the background interference, and the amplitude of deeper reflections is also smaller. In the time–frequency spectrum, point A in Fig. 8.26b corresponds to the reflection from the top plate of the cavity beneath the pile bottom. Although there is no noticeable increase in amplitude energy in the time domain, the frequency and energy peak at point A in the time–frequency spectrum are stronger than the first frequency peak, indicating a clear reflection from the cavity beneath the pile bottom. After point A, the amplitude energy and frequency peak gradually decrease until point B, where another frequency peak appears, corresponding to the reflection from the cavity's bottom plate. In Fig. 8.26d, the frequency peak and its amplitude energy at point A do not show a significant decrease compared to the previous frequency peak, indicating the arrival of a reflection from the pile bottom. Point B represents the reflection from the cavity's bottom plate. In Fig. 8.26f, the interruption in the frequency peak at point

A is caused by the phase interference between the reflection from the cavity beneath the pile bottom and the bottom surface wave of the pile hole. This phase interference manifests as amplitude cancelation in the time domain, making it more difficult to extract effective reflections in the time domain. However, the absence of frequency peaks in the time–frequency spectrum is a strong indicator of the cavity's reflection. Point B clearly reveals the reflection from the cavity's bottom plate. In Fig. 8.26h and j, both points A and B exhibit a sudden increase in frequency peaks, indicating the reflections from the top and bottom plates of the cavity beneath the pile bottom.

From the analysis of cavities at different depths, it can be observed that points A and B appear consecutively with a similar time interval, indicating the accuracy of using the generalized S-transform for detecting cavity reflections. When the cavity is shallow, the detection signal is highly affected by the bottom surface wave in the time domain. However, in the time–frequency domain of the generalized S-transform, the arrival of cavity reflection can be determined based on the absence or sudden increase of frequency peaks. By using Eqs. 6.45 to 6.49, the depth of the cavity top plate corresponding to point A can be quantitatively calculated. After performing calculations with a denser interval, a comparison graph of identification and set values is obtained as shown in Fig. 8.27. The identification of the cavity top plate is generally consistent with the set values, with errors mainly arising from the offset distance and the curved reflection surface of the cavity top plate.

(3) **Varying Sizes of Karst Voids**

Figure 8.28 shows the detection signals and their corresponding time–frequency spectra for cavities of different sizes. The depth of the cavity top plate is 4 m, and the cavity radii are 0.2, 0.6, 1, 1.4, and 1.8 m, respectively. Figure 8.28a, c, e, g, and i represents the detection signals in the time domain. Although the depth of the cavity top plate is the same, there are slight differences in the arrival time of the top plate's reflection wave, and it is difficult to determine the exact arrival time from the time-domain signals. Point A in Fig. 8.28a, c, e, g, and i can be identified by the increase in frequency peaks or amplitude energy. From the corresponding time–frequency spectra, it can be observed that the reflection points A of the cavity top plate, with different sizes, have slight differences in arrival time, where larger cavity radii result in earlier occurrences of point A. This is because larger cavities have shallower true reflection points on the curved top plate, leading to slight differences in the arrival time of the reflection wave at the same depth. As for the reflection of the cavity bottom plate, overall, the deeper the bottom plate of the cavity, the smaller the amplitude of the reflection wave, as more attenuation occurs in larger cavities. In the last scenario shown, for the cavity with a radius of 1.8 m, it is challenging to identify the bottom plate reflection in the time-domain signal of Fig. 8.28i. However, the weak bottom plate reflection can still be well recognized in the time–frequency spectrum of Fig. 8.28j.

From the analysis of different cavity sizes, for larger cavities, the depth of the cavity bottom is greater, and the reflected signal in the time domain is relatively weak. It is challenging to identify it directly in the time domain, but the time–frequency domain provided by the generalized S-transform offers better indications

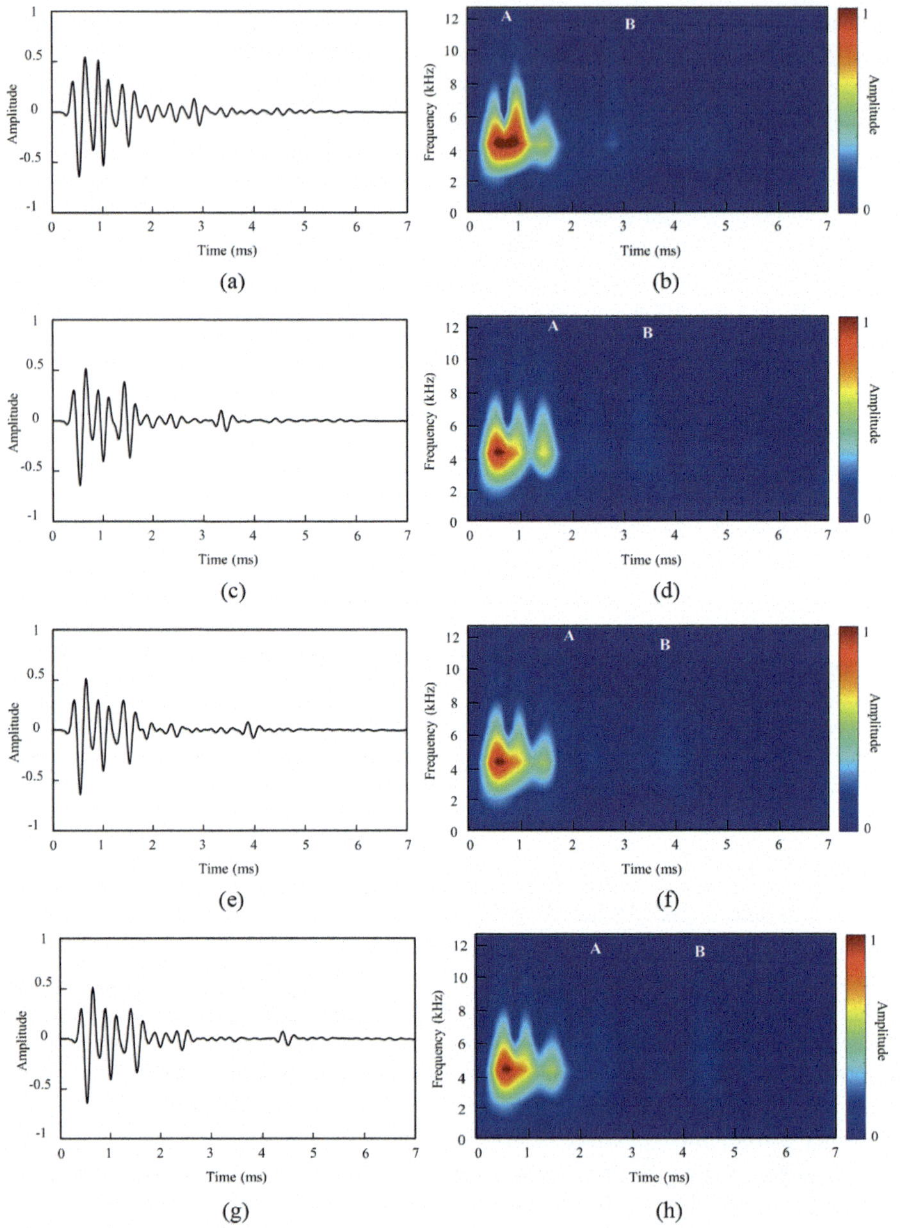


**Fig. 8.26** Analysis of the generalized S-transform for detection signals of cavities at different depths: **a**, **c**, **e**, **g**, and **i** correspond to detection signals with cavity top plate depths of 1, 2, 3, 4, and 5 m, respectively; **b**, **d**, **f**, **h**, and **j** show their corresponding time–frequency spectra based on the generalized S-transform

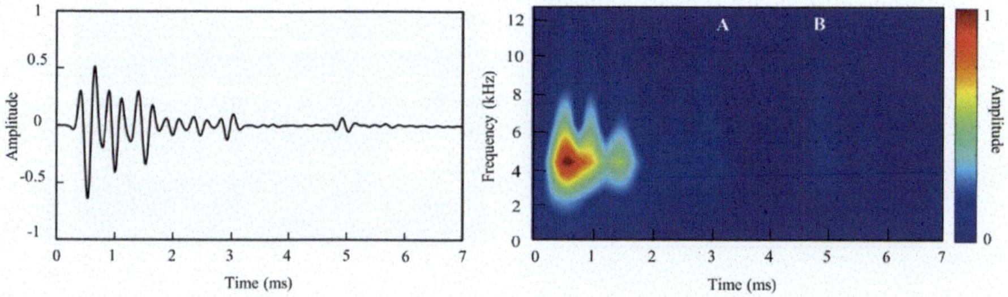


(i) (j)

**Fig. 8.26** (continued)

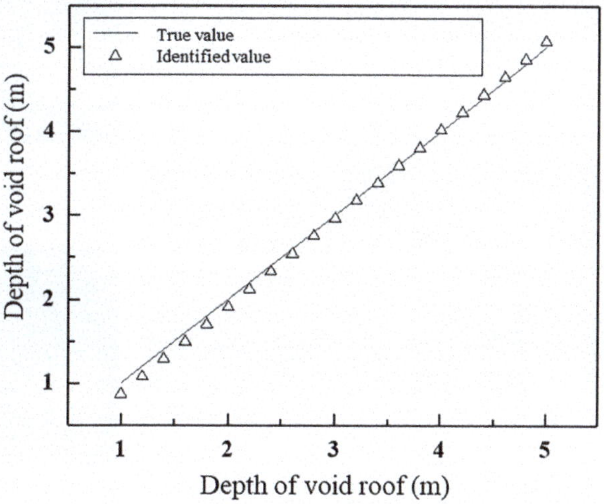


**Fig. 8.27** Comparison between the identified values and set values of the generalized S-transform for the different depths of the cavities roof

of the reflections. Using Formula 8.49, the time difference between points A and B can be quantitatively calculated, and based on this, the radius of the cavity can be estimated. By incrementing the radius with smaller intervals, a comparison between the estimated and set values can be obtained, as shown in Fig. 8.29. The estimated cavity radius generally matches the set value, with errors mainly occurring for larger cavity radii, where the reflected energy is weaker due to greater depth.

(4) **Varying Horizontal Positions of Karst Voids**

In numerical studies in two dimensions, the horizontal position and depth of the cavity can be determined by analyzing the detection signals from two different transmitter–receiver combinations. In realistic three-dimensional conditions, this solution process requires three sets of detection signals from different transmitter–receiver combinations. Figure 8.30 shows the detection signals for a cavity with a top depth of 4 m, a horizontal distance of 1 m from the center axis of the pile bottom, and a radius of 1 m. Two sets of detection signals are shown, where the transmitter is

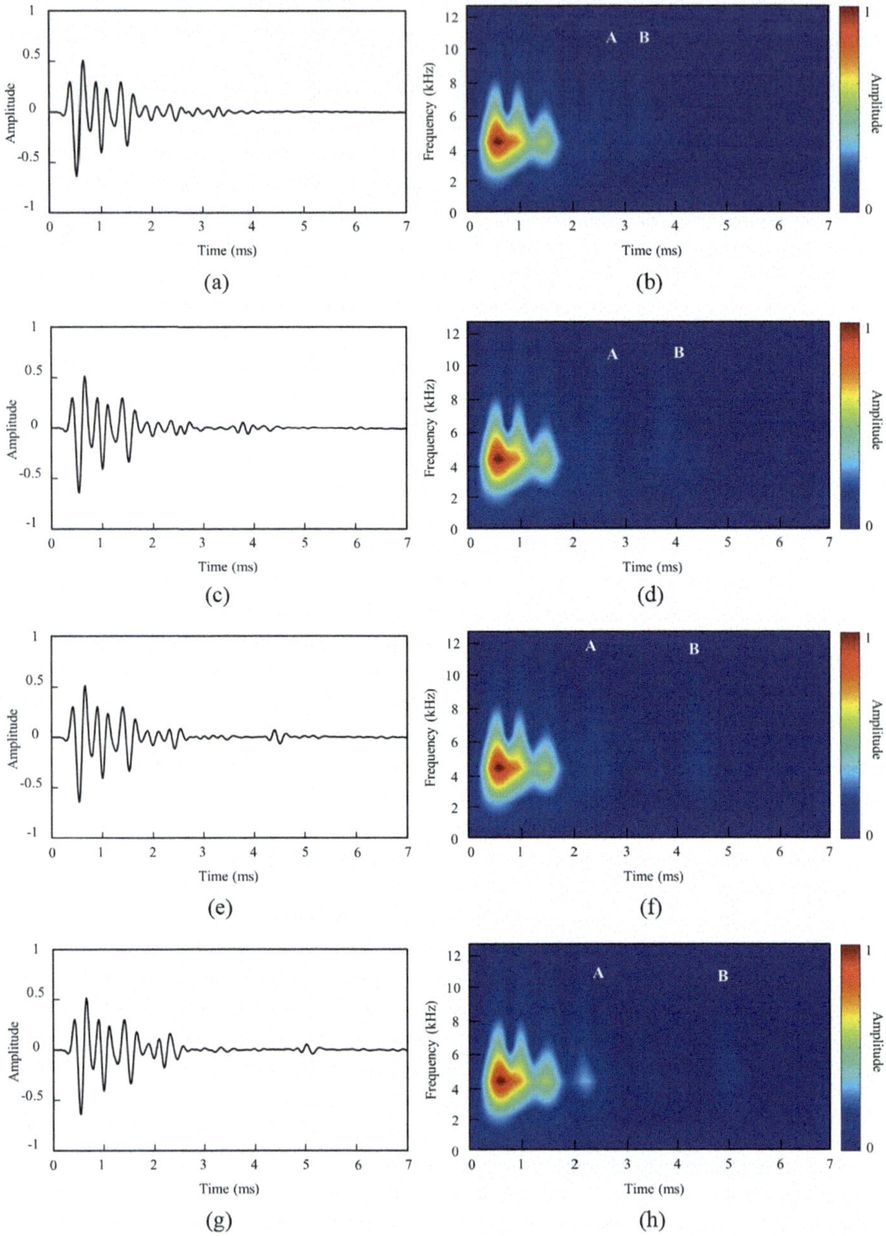


**Fig. 8.28** Analysis of the generalized S-transform for detection signals of different cavity sizes. **a**, **c**, **e**, **g**, and **i** represents the detection signals for cavity radii of 0.2, 0.6, 1, 1.4, and 1.8 m, respectively. **b**, **d**, **f**, **h**, and **j** shows their corresponding time–frequency spectra using the generalized S-transform

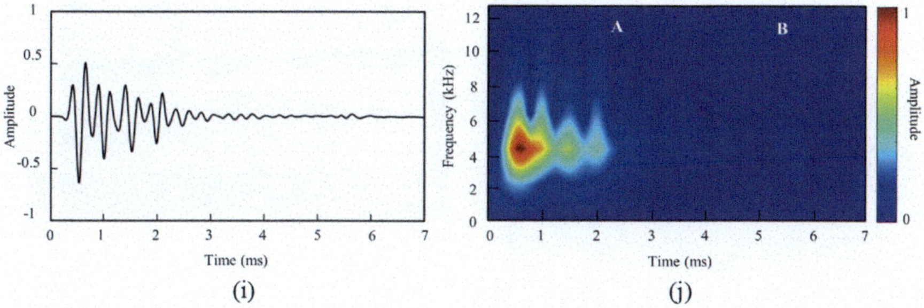


**Fig. 8.28** (continued)

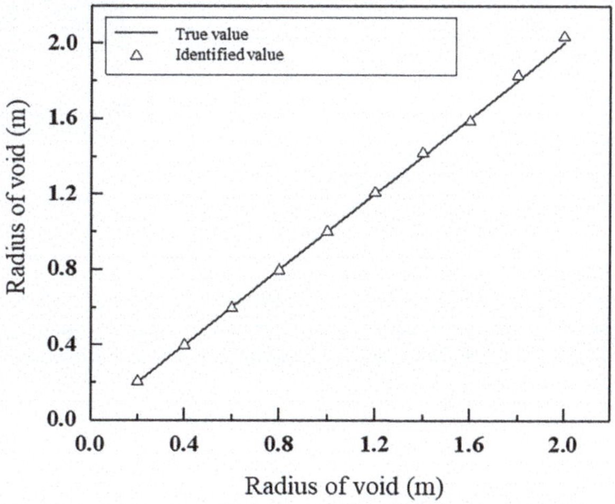


**Fig. 8.29** Comparison between recognized values and set values of generalized S-transform for different cavity sizes

placed in the middle and the receivers are positioned 0.4 m to the left and right sides of the transmitter. Figure 8.30a and b depicts the detection signals recorded by the distant receiver transducer and their corresponding generalized S-transform time–frequency spectra. Points A and B exhibit characteristics of increased frequency peaks and amplitude, corresponding to the arrival times of the cavity's top and bottom reflections, respectively. Figure 8.30c and d displays the detection signals recorded by the near receiver transducer and their generalized S-transform time–frequency spectra. Points A and B also correspond to the arrivals of the cavity's top and bottom reflections. It can be observed that the reflections recorded by the near receiver occur earlier than those recorded by the distant receiver, indicating the resolution capability of the generalized S-transform. The anomalous points in the time–frequency spectra can distinguish the horizontal position difference of the cavity (in three dimensions, this corresponds to the cavity's orientation).

Based on Eq. 8.48, the specific horizontal position can be calculated. In this study, the horizontal positions were calculated from the exact center of the pile bottom to a distance of 5 m away from the center axis of the pile bottom, with a horizontal interval

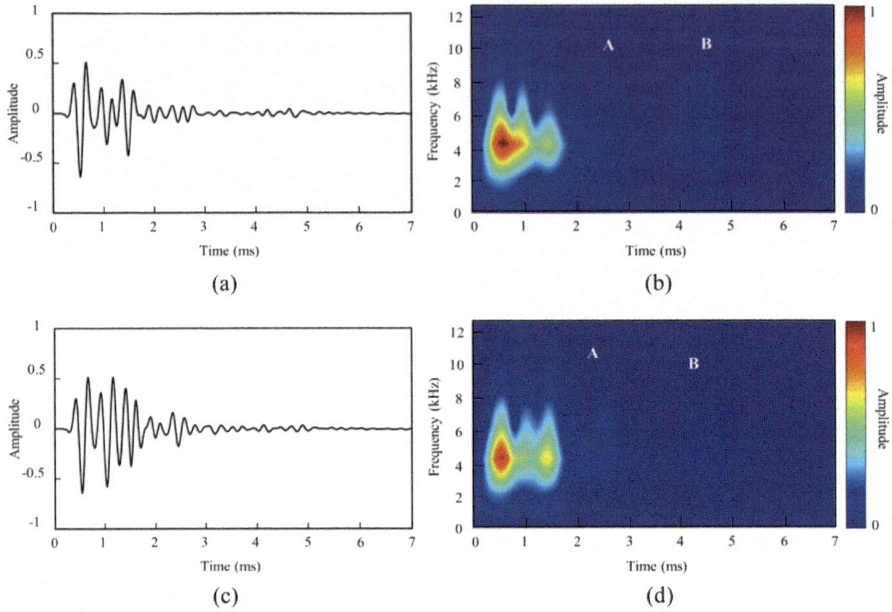


**Fig. 8.30** Analysis of generalized S-transform for detecting signals of different cavity horizontal positions: **a** and **b** show the detection signals and their corresponding generalized S-transform time–frequency spectra recorded by the far-end receiver transducer, while **c** and **d** show the detection signals and their corresponding generalized S-transform time–frequency spectra recorded by the near-end receiver transducer

of 0.2 m for each case. The comparison between the recognized and set values for the horizontal position is shown in Fig. 8.31. The recognized values for the horizontal position of the cavity align closely with the set values. However, there is some error in recognizing the horizontal position of cavities that are far from the center axis of the pile. This might be due to the complex process where some of the reflected waves from cavities located at larger horizontal distances might first arrive at the bottom or sidewall of the pile and then be recorded by the receiver transducer after transmission and waveform conversion. This complex process introduces some interference in identifying the arrival time of the reflected waves. In practical scenarios, when the cavity beneath the pile bottom is located far from the pile, its influence on the bearing capacity of the pile is generally negligible and can be disregarded.

(5) **Voids and Fracture Model**

Figure 8.32 shows a schematic diagram of the model considering the combination of cavities and fractures at the bottom of the pile. The influence of fractures on the recognition of pile-bottom cavities is analyzed through the combined model. In this section, five models are computed, including different depths of fractures and a combination model with a cavity at the same depth. The diameter of the pile

**Fig. 8.31** Comparison between recognized and set values for different horizontal positions of cavities based on generalized S-transform

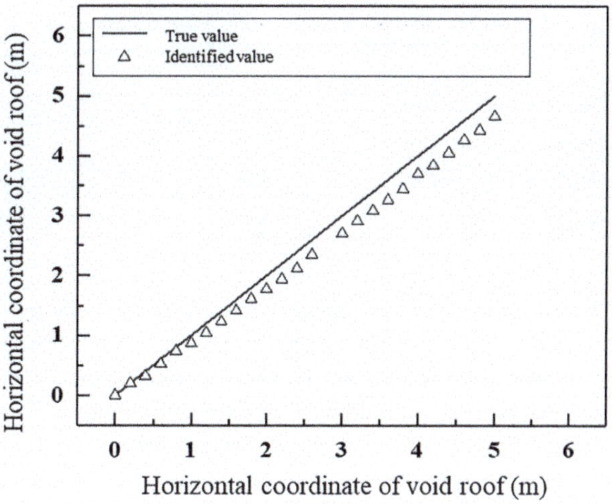


hole is 1.6 m for all models. The geometric parameters of the models are shown in Table 8.10, where the depth of the fractures varies from 1.0 to 3.0 m. The thickness of the fractures, the depth of the cavity top plate, and the radius of the cavity remain unchanged.

Figure 8.33a, c, e, g, and i shows the original detection signals of the five combination models. Due to the presence of fractures, there are multiple reflection interfaces at the bottom of the pile, including multiple reflected surface waves from the pile bottom and reflected waves from the top and bottom of the fractures and cavities.

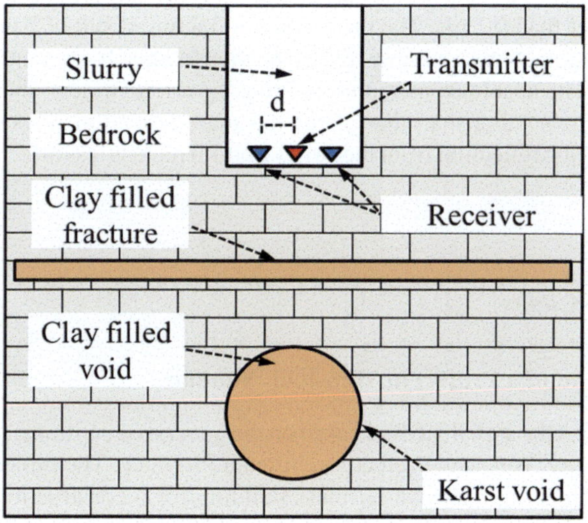


**Fig. 8.32** Combination model of pile-bottom cavities and fractures

**Table 8.10** Geometric parameters of the combination model of pile-bottom cavities and fractures

| Model number | 1 | 2 | 3 | 4 | 5 |
|---|---|---|---|---|---|
| Depth of fracture (m) | 1.0 | 1.5 | 2.0 | 2.5 | 3.0 |
| Thickness of fracture (m) | 0.5 | 0.5 | 0.5 | 0.5 | 0.5 |
| Depth of void roof (m) | 4.0 | 4.0 | 4.0 | 4.0 | 4.0 |
| Radius of void (m) | 1.0 | 1.0 | 1.0 | 1.0 | 1.0 |

The multiple reflected surface waves from the bottom of the pile should exhibit equal intervals of attenuation, so any anomalies observed in addition to these intervals can be attributed to reflections from other interfaces at the bottom of the pile. However, this recognition is challenging in the time domain. The results of the generalized S-transform applied to the signals are shown in Fig. 8.33b, d, f, h, and j. The reflected waves from fractures at different depths are evident at 0.5–1 ms in Fig. 8.33b, at 1 ms in Fig. 8.33d, at 1.0 ms to 1.5 ms in Fig. 8.33f, at 1.5 ms in Fig. 8.33h, and at 1.5 ms to 2 ms in Fig. 8.33j, exhibiting strong energy in the reflected waves. The reflected waves from horizontally oriented fractures have stronger energy compared to the point-like reflections from cavities.

Points A and B in Fig. 8.33b, d, f, h, j represent frequency peaks caused by the reflections of the top and bottom plates of the cavity. These frequency peaks are difficult to identify in the corresponding time-domain signals. However, through the generalized S-transform, clear frequency peaks are observed in the time–frequency spectra. The arrival times of the top and bottom plate reflections, calculated from points A and B, respectively, are shown in Table 8.11. The theoretical values are obtained by subtracting the models with cavities from the models without cavities. The results of the identification show that the absolute errors of most reflection arrival times are smaller than 0.1 ms. The model with a fracture depth of 2 m has relatively larger identification errors, possibly due to the interference from multiple reflections of the fracture. The absolute error in identifying the arrival time of the cavity bottom plate reflection is smaller than that of the top plate reflection. This is because the identification of the reflection from the cavity bottom plate is less affected by multiple reflections.

### *8.4.3 Field Case*

#### (1) **Background of a Project in Wenshan, Yunnan**

This paper applies the signal method based on the generalized S-transform to a project in Maguan County, Wenshan Prefecture, Yunnan Province. The project involves the construction of a comprehensive business building for a construction credit union. It is a framed structure with a total floor area of 809.18 square meters. During the construction of the foundation piles, it was found that there were multiple layers of

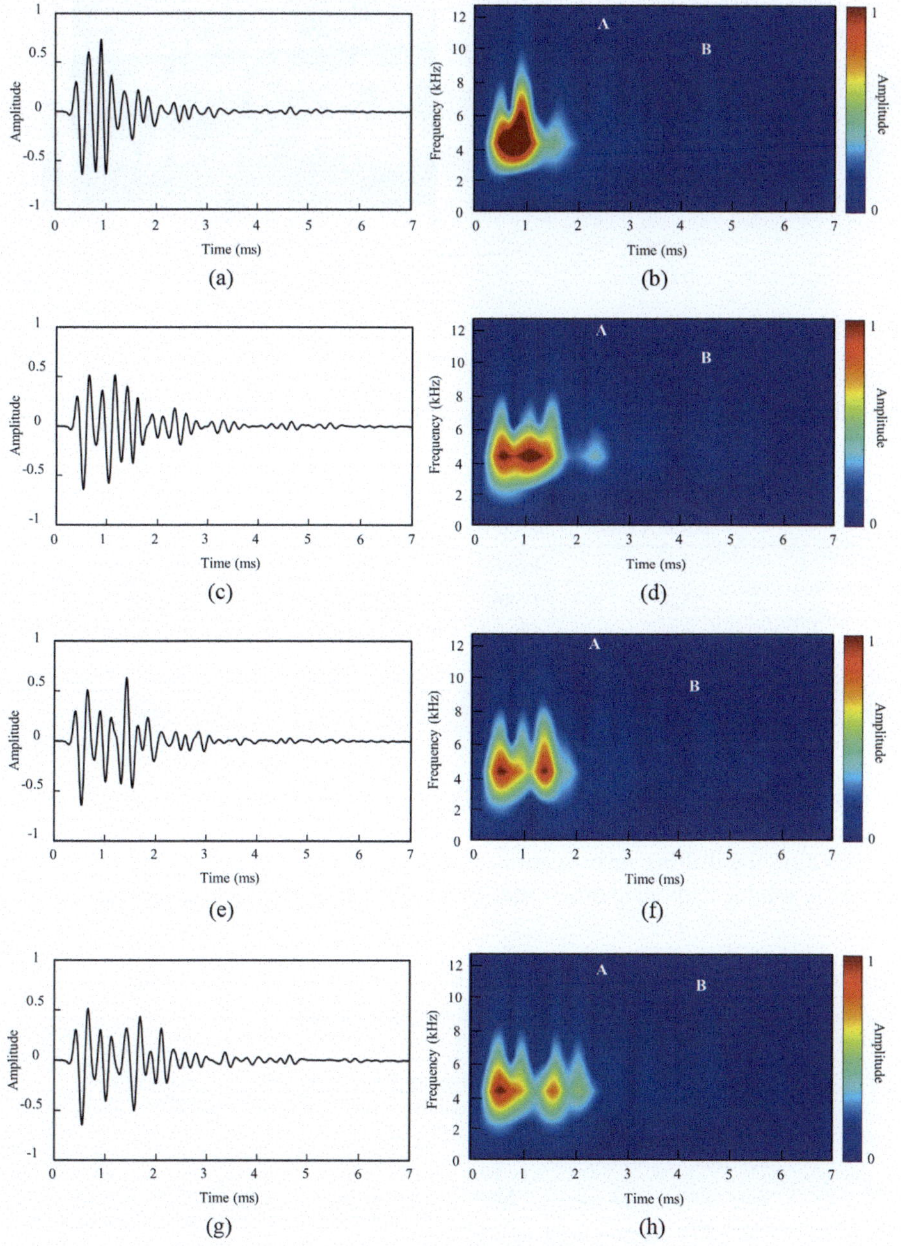


**Fig. 8.33** Analysis of the generalized S-transform applied to the detection signals of different combinations of cavities and fractures. **a**, **c**, **e**, **g**, and **i** correspond to the detection signals when the horizontal fractures are located at 1.0, 1.5, 2.0, 2.5, and 3.0 m, respectively. **b**, **d**, **f**, **h**, and **j** represent their corresponding time–frequency spectra obtained through the generalized S-transform

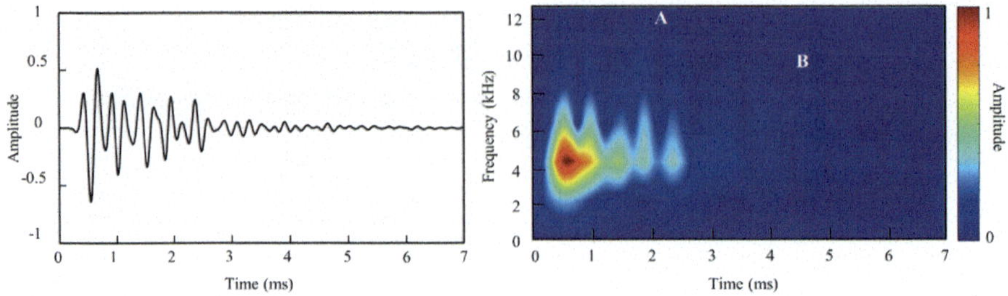


(i) (j)

**Fig. 8.33** (continued)

**Table 8.11** Calculation results and errors of the arrival times for the reflections from the top and bottom of the cavity beneath the pile bottom

| Depth of fracture (m) | Cavity top: Arrival time of point A (ms) | Cavity bottom: Arrival time of point B (ms) | Absolute error of arrival time of point A (ms) | Absolute error of arrival time of point B (ms) |
|---|---|---|---|---|
| 1.0 | 2.300 | 4.686 | 0.072 | 0.026 |
| 1.5 | 2.324 | 4.686 | 0.048 | 0.026 |
| 2.0 | 2.254 | 4.694 | 0.118 | 0.034 |
| 2.5 | 2.342 | 4.682 | 0.030 | 0.022 |
| 3.0 | 2.292 | 4.702 | 0.080 | 0.042 |

floating stones or boulders in the alluvial deposit layer. The geological layers at the site, from top to bottom, are described in Table 8.12.

Based on the drilling investigation, it is revealed that the underlying bedrock at the site is soluble carbonate limestone, with karst landforms mainly characterized by

**Table 8.12** Characteristics of the soil and rock layers at a construction site in Yunnan

| Formation period | Soil layer | Profile image | Soil layer characteristics |
|---|---|---|---|
| $Q_4^{al}$ | Fourth series fluvial deposits | 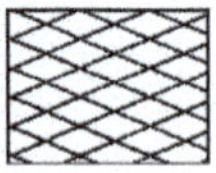 | Mainly soft plastic silty clay, with uniform soil composition, average thickness of 1.6 m |
| $Q_p^{al}$ | Silty clay layer with angular gravel |  | Mainly composed of silty clay, with gravel composed of quartzite and sandy rock. Some areas contain multiple layers of large stones. The average thickness is 10.1 m |
| $\in_3$b | Broken limestone layer | 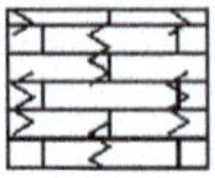 | The limestone is moderately weathered with cryptocrystalline structure. It has a blocky structure and is a hard rock. The rock mass is fractured, and its quality grade is Class IV |

caves, grooves, and small erosional holes. The depth and scale of karst development vary within the exploration depth range, with burial depths ranging from 19.50 to 23.40 m. The depth of the karst cavities at the pile bottoms ranges from 4.60 to 7.50 m, with cavity heights ranging from 0.80 to 1.50 m.

During the advanced drilling for the inspection piles, a total of 18 piles were tested, and five of them (A-2, A-5, B-2, B-5, and C-3) were found to have karst cavities at their bottoms, accounting for 27.78% of the tested piles. The cavity detection rate was 5.33%. The total length of limestone encountered was 622.6 m, with a cumulative height of karst cavities reaching 5.30 m. The linear karst rate was 0.85%. The karst development exhibited a weak regularity, indicating a site with relatively weak karst development. Some local dissolution phenomena were observed in fractured and intact limestone layers. Groundwater is present at the site within the exploration depth range, and under the influence of groundwater, further karst development is anticipated.

Considering the aforementioned karst characteristics, the proposed site is classified as a site with relatively weak karst development and a weak regularity in karst development. The presence of karst cavities can affect the stability of the foundation. For safety reasons, the detection of karst cavities at the pile bottoms was conducted using sonic methods for all piles at the site.

(2) **Analysis of Generalized S-transform for Detecting Karst Cavities Beneath Pile Bottoms**

During on-site detection, it is required that the construction party cleans the residual debris at the bottom of the pile. If there is no water at the pile bottom, water should be injected to ensure that the water depth at the pile bottom is not less than 40 cm. Then, the sonar probe is inserted into the pile bottom, and the probe position is adjusted based on the tilt angle displayed on the instrument, ensuring that the probe's horizontal angle is less than 5° before starting the detection.

Figure 8.34 shows the detection wave profiles obtained from the pile foundation hole labeled as D2. The pile length is 12.7 m, and data were collected from eight different azimuth angles, with azimuth angles of channels 1 to 8 being 37, 66, 127, 156, 217, 246, 307, and 336°, respectively. From the original profile, the arrival time of the direct wave can be observed at point A, and there are noticeable reflections at points B and D. The reflection at point C is not very prominent, and in channel 2, the reflection is not clear. To further analyze the arrival time of reflections and the nature of reflection waves more clearly, the generalized S-transform method is employed.

Figure 8.35 shows the time–frequency spectra of the eight channel signals obtained from the sonar detection of the pile base hole. Overall, the time–frequency spectra of the sonar detection are similar to the combined conditions of fractures and cavities (Fig. 8.33). In Fig. 8.35a and b, point A corresponds to the direct wave, point B corresponds to the attenuation of the multiple surface waves at the pile base, and points C and D exhibit frequency peaks, representing the reflections of the cavity's top and bottom surfaces, respectively. In Fig. 8.35c and d, point A corresponds to the direct wave, and point B shows a significant enhancement in signal energy. Based on the numerical simulation results of the combined model of fractures and cavities, point B corresponds to the reflection of the linear fracture at the pile base, which has stronger wave energy compared to the reflection from cavity. Point C, with no significant decay in frequency peak, is determined as the emission from the cavity's top surface. Compared to the cavity reflection in Channels 1 and 2 (points C in Fig. 8.35a and b), multiple noise interferences appear after point C in Channels 3 and 4, which can be attributed to the multiple reflections caused by the combination of fractures and cavities. Point D corresponds to the reflection from the cavity's bottom surface. Fig. 8.35e and f show the time–frequency spectra of Channels 5 and 6, which are similar to Channels 3 and 4. Point B represents the reflection from the fracture at the pile base, but with an earlier arrival time and weaker wave energy compared to Channels 3 and 4. Points C and D correspond to the reflections from the cavity's top and bottom surfaces, respectively. The time–frequency spectra of the detection signals in Channels 7 and 8 are shown in Fig. 8.35g and h. Point B, representing the reflection from the fracture, has similar arrival times as in Channels 5 and 6. The reflection point C from the cavity's top surface is not clear, but the reflection point D from the bottom surface is still observable.

Based on the rock lithology and integrity at the site, the estimated longitudinal wave velocity of the rock is approximately 3000 m/s. According to the time–frequency analysis results, it is believed that there is a rock fracture at a depth of approximately 1.5 m on the southwest side of the pile base, and there is a cavity's top surface at a depth of 2.7 m below the pile base. If the cavity is filled with clay, its diameter is estimated to be approximately 1 m. The cavity does not exhibit a clear directional pattern. It is recommended to grout after penetrating the cavity's top surface at the pile base. Subsequent on-site construction feedback confirmed the

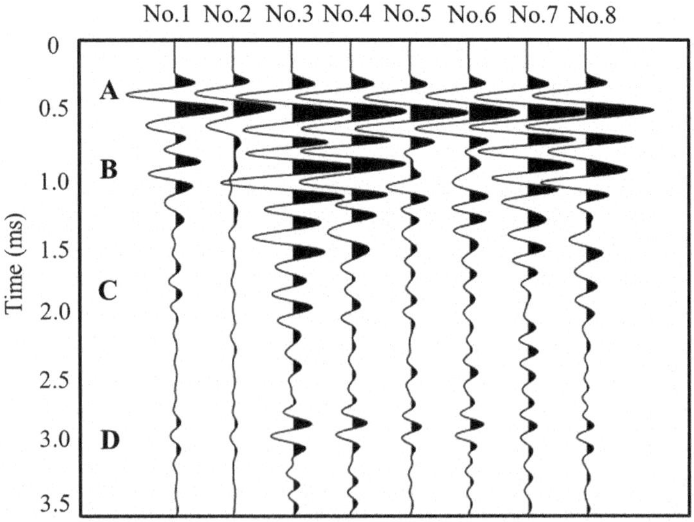


**Fig. 8.34** Wave profile of pile hole sonar detection

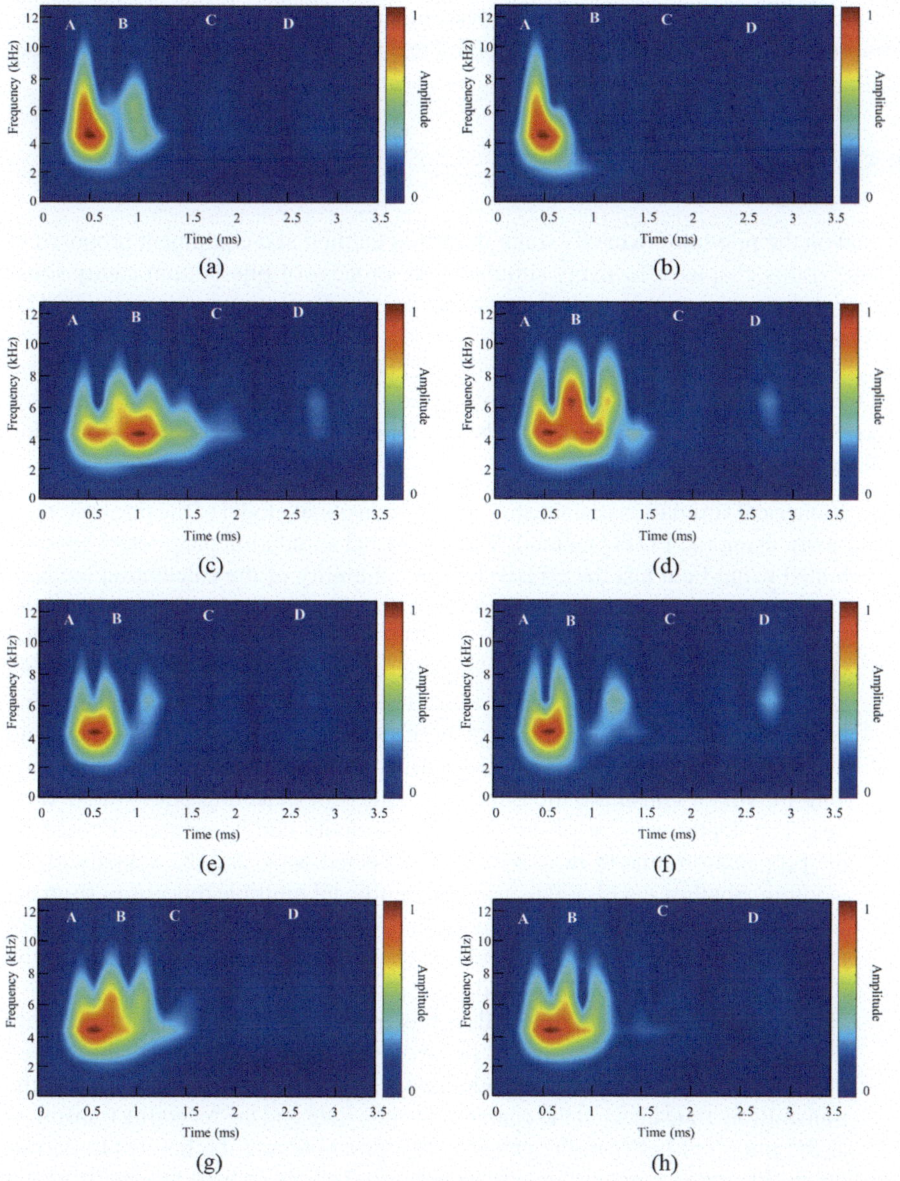


**Fig. 8.35** Time–frequency analysis of the generalized S-transform for the eight channels. **a** 37 degrees in channel 1, **b** 66 degrees in channel 2, **c** 127 degrees in channel 3, **d** 156 degrees in channel 4, **e** 217 degrees in channel 5, **f** 246 degrees in channel 6, **g** 307 degrees in channel 7, **h** 336 degrees in channel 8

validity of the detection results and analysis methods, as evident from the noticeable grouting leakage occurring at a depth of 3.1 m during the advanced drilling process.

## 8.5 Summary

Based on the pile-bottom cavity sonar detection method and equipment proposed in Chap. 7, this chapter effectively simulated the process of pile-bottom cavity sonar detection using three-dimensional cross-grid finite-difference time-domain (FDTD) method and revealed the wavefield components and detection mechanisms. Based on this, qualitative rapid analysis and single-channel signal analysis methods based on the generalized S-transform are proposed. These signal processing methods were validated using numerical signals and field signals, leading to the following conclusions。

(1) Numerical simulations establish a three-dimensional model of the pile hole-rock cavity using the FDTD method. The numerical simulation parameters are calibrated using field test data, verifying the reliability of the calculation method. The numerical simulation work reveals that the waveforms recorded by the transducer at the pile hole bottom mainly consist of two parts: one part is the waves propagating horizontally along the pile hole bottom and their multiple reflections on the pile wall, which constitute the main interference in detecting pile-bottom cavities; the other part is the waves propagating downward into the bedrock and reflecting at geological structural interfaces beneath the pile bottom, which constitute the effective signals for detecting pile-bottom cavities. The depth and size of the pile-bottom cavity can be estimated based on the propagation time of the cavity's PP-reflected wave and the velocity of the medium. Shallow cavities may produce significant multiple reflections from the cavity's top and strong oscillations in the frequency spectrum. The azimuth of the pile-bottom cavity can be calculated by analyzing the time difference of the reflected waves recorded by transducers at different azimuth angles.
(2) The direct waves and multiple reflections from the pile hole wall are the main interferences in the detection of pile-bottom cavities. To reduce the impact of multiple reflections from the pile hole wall, it is recommended to place the transmitting transducer at the center of the pile hole and the receiving transducer at the midpoint between the center of the pile hole and the hole wall. At this point, the signals from multiple reflections on the hole wall are symmetrical and easier to identify.
(3) When the cavity is completely filled with a low-velocity medium, the cavity exhibits wave focusing effects. The energy of the reflected waves from the cavity's bottom is theoretically stronger than that from the cavity's top. The amplitude of the reflected waves increases with the impedance difference between the bedrock and the cavity filling material. However, regardless of whether the cavity is filled with a medium or not, the impedance of air, water,

clay, and other substances inside the cavity is much smaller than that of the bedrock on the cavity's top, resulting in a phase reversal of the superimposed reflected waves compared to the incident waves, leading to strong signal consistency.

(4) Pile-bottom cavity detection does not rely on advance drilling and is a fast and nondestructive testing method that can be applied to routine inspections of every pile. Field detection of a single pile can be completed within 10 min, and pile-bottom cavity tracking detection can be performed at any time during the drilling process. It brings great convenience and operability to the quality inspection of the pile-bottom rock, making this technology easy to popularize. The pile-bottom cavity sonar detection method proposed in this study has been validated at various engineering sites in Nanning, Guangxi, Anshun, Guizhou, Wenshan, Yunnan, etc. The detected signal characteristics are distinct, and the results have been verified through advance drilling.

(5) This section presents an analysis method for pile-bottom cavities and fissures based on the generalized S-transform. The feasibility of identifying pile-bottom cavities and fissures using the generalized S-transform is validated through numerical simulations and field data analysis. The multiple reflections from the pile hole bottom wall exhibit regular frequency and amplitude attenuation in the time–frequency spectrum. However, the effective reflected waves appear as amplified amplitudes and frequency peaks in the time–frequency spectrum. The arrival times of these peaks can be analyzed to determine the arrival time of the reflected waves and thereby determine the depth of the pile-bottom karst geological structure.

(6) The analysis based on the generalized S-transform focuses on single-channel signal time–frequency analysis, which has strict requirements for the position of the transducer and works best when the transmit–receive distance is 0.5 times the pile radius. The analysis is still susceptible to irregular multiple reflections from the pile hole bottom wall. The information provided by single-channel signal analysis is limited. In the following Chap. 7, a multi-channel joint analysis will be considered to propose more powerful signal processing and automatic reflected wave extraction methods.

## References

Collino F, Tsogka C (2001) Application of the perfectly matched absorbing layer model to the linear elastodynamic problem in anisotropic heterogeneous media. Geophysics 66(1):294–304

Dong L, Ma ZT, Cao J (2000a) Stability of the staggered-grid high-order difference method for FIRST-order elastic wave equation. Chinese J Geophys 43(6):904–913

Dong L, Ma ZT, Cao J (2000b) Staggered-grid high-order difference method for the first-order elastic wave equations. Chinese J Geophys 43(3):441–449

Dong L, Huang C, Liu Y (2010) Numerical simulation of seismic wave propagation in cave carbonate reservoir. Geophys Prospect Petrol 49(2):121–124

Hosokawa A (2011) Numerical investigation of ultrasound refraction caused by oblique orientation of trabecular network in cancellous bone. IEEE Trans Ultrason Ferroelectr Freq Control 58(7):1389–1396

Hosokawa A (2010) Effect of porosity distribution in the propagation direction on ultrasound waves through cancellous bone. IEEE Trans Ultrason Ferroelect Freq Control 57(6):1320–1328

Lu ZT, Wang ZL, Liu DJ (2013) Study on low strain integrity testing of pipe pile using the elastodynamic finite integration technique. Int J Numer Anal Meth Geomech 37(5):536–550

Marcinkovich C, Olsen K (2003) On the implementation of perfectly matched layers in a three-dimensional fourth-order velocity-stress finite difference scheme. J Geophys Res Solid Earth (1978–2012) 127–143

Pinnegar CR, Mansinha L (2004) Time-local Fourier analysis with a scalable, phase-modulated analyzing function: the S-transform with a complex window[J]. Signal Process 84(7):1167–1176

Saenger EH, Bohlen T (2002) Finite-difference modeling of viscoelastic and anisotropic wave propagation using the rotated staggered grid. Seg Tech Progr Expanded Abstr 21(1):2478

Schubert F, Peiffer A, Köhler B et al (1998) The elastodynamic finite integration technique for waves in cylindrical geometries. J Acoust Soc Am 104(104):2604–2614

Shi Z, Liu L, Ming P et al (2016) A sonar detection technology for karst cavities under bored piles and its application. Chinese J Rock Mech Eng

# Chapter 9
# Advanced Pile Hole Sonar Detection System

## 9.1 Prologue

In Chap. 8, we introduced a pile hole sonar technique that utilizes a single receiver-transmitter pair positioned at the bottom of the pile hole. This approach aims to image the subsurface beneath the pile hole by analyzing P-wave seismic reflection data. Previous studies (Shi et al. 2016, 2018; Liu et al. 2018) have identified challenges in this method, particularly the interference caused by surface-wave multiples propagating along the pile hole. These multiples can obscure the weaker reflections from voids beneath the pile hole. To address this issue, we draw inspiration from surface seismic acquisition techniques that utilize geophone arrays to suppress interfering wave modes and enhance the signal-to-noise ratio (Craig and Genter 2006; Wu and Huang 2019). We adapt this concept to our pile hole sonar imaging approach in order to extract and enhance the faint reflections associated with karst voids.

In this chapter, we introduce a novel sonar method for detecting karst cavities beneath piles. Our approach involves utilizing two receivers positioned at different distances from the transmitter, allowing us to capture recordings of body wave reflections from voids and distinguish them from interfering surface-wave multiples along the pile hole. To achieve this, we employ instantaneous phase analysis of the recorded signals, which has been proven to be effective in discerning different waveform characteristics (Taner et al. 1979). We utilize the Hilbert-Huang transform (HHT), which combines Empirical Mode Decomposition (EMD) (Huang et al. 1998) and the Hilbert transform (HT) (Taner et al. 1979), to calculate the instantaneous phase and frequency of the nonstationary and nonlinear data. To overcome the limitations of EMD, we utilize Ensemble Empirical Mode Decomposition (EEMD) (Wu and Huang 2009) and complete Ensemble Empirical Mode Decomposition (CEEMD) (Torres et al. 2011), which have been widely applied in seismic studies (Battista et al. 2007; Fomel 2013; Honório et al. 2017). Additionally, we introduce a new index called Instantaneous Phase Difference Intensity (IPDI) to quantify the similarity of instantaneous phase between two traces and effectively differentiate P-wave



L. Liu et al., *Detection of Karst Voids at Deep Pile Foundation*,
https://doi.org/10.1007/978-981-97-5834-0_9

arrivals from surface-wave arrivals. To validate our method, we present results from modeling experiments and field test using prototype instrumentation.

## 9.2 Two-Receiver Pile Hole Acquisition

To distinguish weak cavity reflections from interfering surface-wave modes at the bottom of the pile hole, a new approach is proposed using two receivers positioned at a distance of approximately half the wavelength ($\lambda$) of the surface waves, as shown in Fig. 9.1. The downhole tool configuration includes a transmitter located at the center of the pile hole (Point O in Fig. 9.1) and two receivers (No. 1 receiver at Point A and No. 2 receiver at Point B in Fig. 9.1). The distance between the two receivers and the transmitter can be adjusted based on the pile hole diameter. Numerical simulations are conducted to analyze the signals received at the two receiver locations in models with and without a cavity, as depicted in Fig. 9.2. The presence of a void is not readily apparent in the P-wave reflection (Fig. 9.2a) due to the dominance of surface-wave multiples propagating along the bottom of the pile hole. Simple summation (stacking) of the traces from the two receivers does not effectively mitigate the interference caused by the surface-wave multiples.

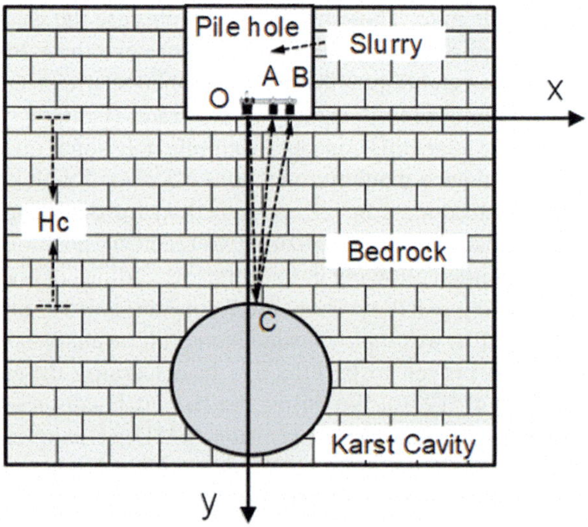


**Fig. 9.1** Schematic cross-sectional view of the novel two-receiver sonar configuration is presented, as shown in Fig. 9.1. The receiver locations are denoted by A and B. For this particular illustration, the pile hole diameter is 1.6 m. The separation between the transmitter and the first receiver (OA) is 0.4 m, while the distance between the transmitter and the second receiver (OB) is 0.6 m. The cavity beneath the pile hole is situated 3 m below the top of the pile hole and has a diameter of 2 m. The material parameters of the slurry, bedrock, and cavity infill correspond to scenario 4, as outlined in Table 9.2

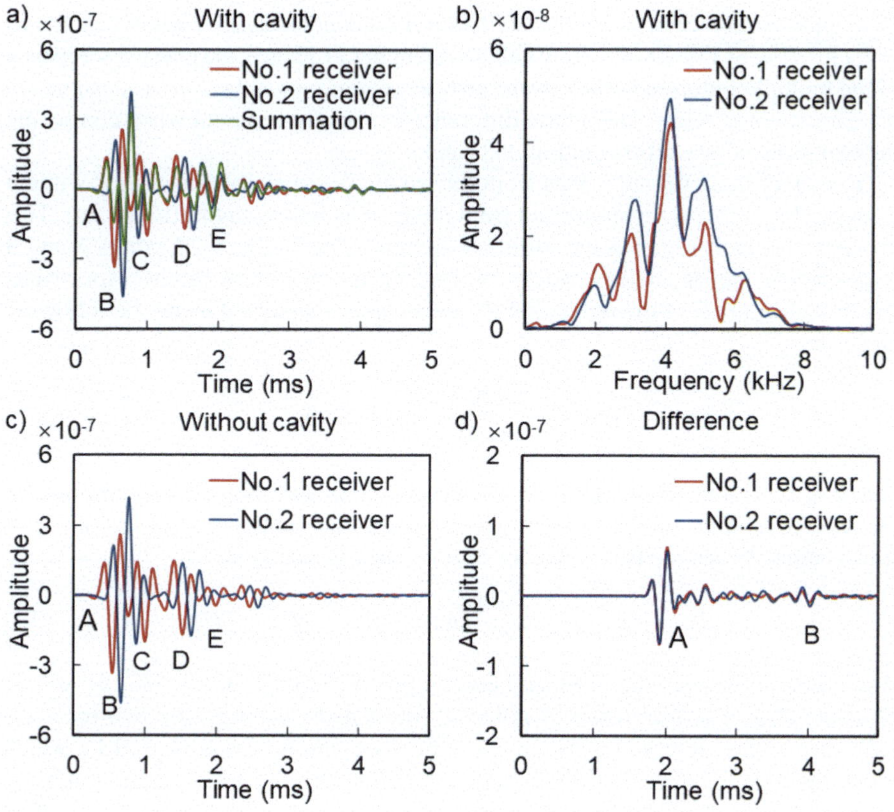


**Fig. 9.2** Numerical simulations were conducted to evaluate the performance of the two-receiver setup depicted in Fig. 9.1. The results are presented as follows: **a** Simulated waveforms with a cavity beneath the pile hole recorded at the two receivers, along with the sum of the two waveforms. **b** Corresponding frequency spectra of the recorded signals. **c** Modeling test signal without a cavity. **d** Reflected wave from the cavity, obtained by subtracting the simulated signal without a cavity (Fig. 9.2c) from the one with a cavity (Fig. 9.2a). In Fig. 9.2a and c, points A, B, C, D, and E are used to identify the direct arrival and the multiple reflected surface waves. In Fig. 9.2d, points A and B indicate reflections from the top and base of the cavity

Points A, B, C, D, and E in Fig. 9.2a and c identify the direct arrival (A) and the multiple reflected surface waves. The wavelength $L_1$ of the multiple reflected surface waves is calculated by:

$$\mathrm{L_1} = \frac{\mathrm{v_r}}{\mathrm{f_m}} = \frac{1713\ \mathrm{m/s}}{4150\ \mathrm{Hz}} = 0.413\ \mathrm{m}, \tag{9.1}$$

where $v$ represents the velocity of the surface waves propagating along the bottom of the pile hole, estimated through analysis of simulated wavefields, and $f_m$ corresponds to the dominant frequency of the signals displayed in Fig. 9.2b. The surface-wave travel path difference between the two receivers is denoted as $\triangle D_1$ equivalent to AB = 0.2 m in Fig. 9.1 (approximately half wavelength $L_1$). Hence, the phase discrepancy of the multiple reflected surface-wave arrivals is approximately $\pm\pi$, as designed in the experimental setup. This phase difference is visually apparent when comparing the two received waveforms in Fig. 9.2a and c.

To extract the reflected waves from the cavity, the modeled waveforms without a cavity (Fig. 9.2c) were subtracted from the models with a cavity (Fig. 9.2a). This process effectively removes the surface-wave multiples. In Fig. 9.2d, points A and B correspond to the reflected PP-waves from the top and base of the cavity, respectively. The wavelength L2 of the reflected PP-waves can be calculated using the following equation:

$$L_2 = \frac{v_p}{f_m} = \frac{4000\ \text{m/s}}{4150\ \text{Hz}} = 0.964\ \text{m}, \tag{9.2}$$

where $v_p$ represents the velocity of P-waves in carbonate bedrock. Assuming that the reflection points on the top of the cavity are nearly coincident, the difference of travel path between the two receivers $\triangle D_2$ can be calculated as:

$$\triangle D_2 = D_{OC} + D_{CA} - (D_{OC} + D_{CB}) = D_{CA} - D_{CB}, \tag{9.3}$$

where $D_{OC}$ represents the distance between points O and C in Fig. 9.1, as well as $D_{CA}$ and $D_{CB}$. The difference of reflected travel paths between the two receivers $\triangle D_2$ approaches zero when the cavity is far from the receivers (Fig. 9.1), which represents (i.e., $\triangle D_2 << L_2$). This characteristic enables the differentiation of weak cavity reflections from multiple reflected surface-wave modes based on their distinct instantaneous phase characteristics.

## 9.3 Instantaneous Phase Difference Intensity Analysis

In this chapter, we introduce a novel method for distinguishing P-wave void reflections from interfering surface waves based on instantaneous phase analysis. The proposed method utilizes Ensemble Empirical Mode Decomposition (EEMD), Hilbert transform (HT), and a newly developed equation to calculate the Instantaneous Phase Difference Intensity (IPDI). The analysis procedure is summarized in Fig. 9.3 and outlined as follows:

(1) Implement the two-receiver configuration as depicted in Fig. 9.1, where the spacing between the two receivers is precisely half the wavelength of the interfering reflected surface waves along the bottom of the pile hole.

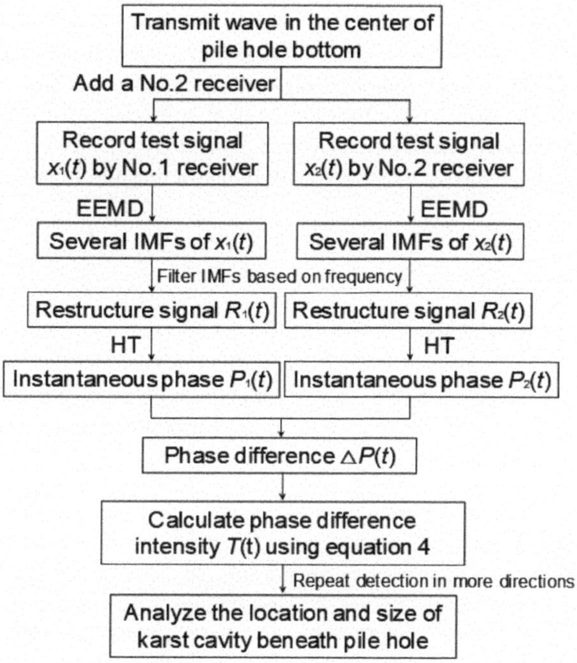


**Fig. 9.3** Flowchart of signal analysis method for void detection based on instantaneous phase analysis

(2) Record signals by No. 1 receiver, denoted as $x_1(t)$, and by No. 2 receiver, denoted as $x_2(t)$ (Fig. 9.1). The recorded signals $x_1(t)$ and $x_2(t)$ are decomposed into several Intrinsic Mode Functions (IMFs) using EEMD. The IMFs possess favorable characteristics for accurate Hilbert transform calculations.

(3) Restructure signals $R_1(t)$ and $R_2(t)$, respectively. IMFs $c_{1i}(t)$ and $c_{2i}(t)$ are selected based on their frequency spectrum. Only the IMFs containing frequencies ranging from 1 to 8 kHz are employed for signal reconstruction.

(4) Compute the instantaneous phase difference between $R_1(t)$ and $R_2(t)$. Utilize HT to calculate the instantaneous phase $P_1(t)$ and $P_2(t)$ of the restructured signals $R_1(t)$ and $R_2(t)$. Obtain the phase difference $\triangle P(t)$ by subtracting one phase from the other.

(5) Calculate the Instantaneous Phase Difference Intensity (IPDI) $I(t)$. The IPDI I(t) is defined as a metric to measure the similarity of the phase between two signals within a specific time period. It is calculated using the following equation:

$$I(t_n) = \int_{t_n - \frac{T}{2}}^{t_n + \frac{T}{2}} ((|\triangle P(t)|)^2 + (2\pi - |\triangle P(t)|)^2 - 2\pi^2)\mathrm{d}t, \tag{9.4}$$

where $t_n$ represents any time point within the signal and T is the signal's period, which is equivalent to the reciprocal of the dominant frequency. The IPDI, $I(t_n)$, reaches a maximum when $t_0$ corresponds to the middle time point of the first period of reflections. By employing Eq. 9.5, the arrival time, $t_a$, of the reflections from a karst cavity can be determined. This arrival time information aids in identifying the location and size of the cavity beneath the pile hole.

$$t_a = t_0 - \frac{T}{2}. \tag{9.5}$$

(6) Repeat steps 1–5 for different receiver azimuth orientations in the pile hole. A minimum of three varying azimuthal detections is recommended to determine the location and size of a karst cavity positioned off-center from the base of the pile ho. This allows for a comprehensive analysis and enhances the accuracy of the results.

The analysis method utilizes Ensemble Empirical Mode Decomposition (EEMD) to achieve two key objectives. Firstly, it enables the application of Hilbert transform (HT) on nonstationary and nonlinear time series data, such as raw detection data from real cases. Secondly, EEMD acts as a noise filter while preserving the crucial phase information. EEMD, an enhanced version of Empirical Mode Decomposition (EMD) developed by Wu and Huang (2009), overcomes limitations present in EMD, such as mode mixing. The EEMD algorithm introduces white noise to the signal in multiple trials, allowing for improved decomposition and analysis of the data.

## 9.4 Numerical Tests of the Instantaneous Phase Difference Intensity Analysis

To demonstrate the IPDI analysis method, we utilize the model signals depicted in Fig. 9.2a. By applying EEMD to these waveforms, we extract several Intrinsic Mode Functions (IMFs). In this analysis, the standard deviation of the added noise is set to 0.1, and the EEMD is performed with an ensemble number of 100. Figure 9.4 showcases the IMFs obtained from the signal recorded by receiver No. 1, along with their corresponding Fourier spectra. It is observed that the first two IMFs primarily represent high-frequency added white noise. The subsequent IMFs (3rd to 6th) correspond to frequencies of 4.150, 4.150, 2.197, and 1.220 kHz, respectively. IMFs with dominant frequencies below 1 kHz can be effectively filtered in the subsequent analysis.

In Fig. 9.5a, the restructured signals from the two receivers are obtained by summing the 3rd to 6th IMFs. The Hilbert transform is then applied to compute the instantaneous phase of these restructured signals. As depicted in Fig. 9.5b, the instantaneous phase exhibits a range between $-\pi$ and $\pi$. Notably, the instantaneous phase of the restructured signal from receiver No. 1 consistently lags behind that of

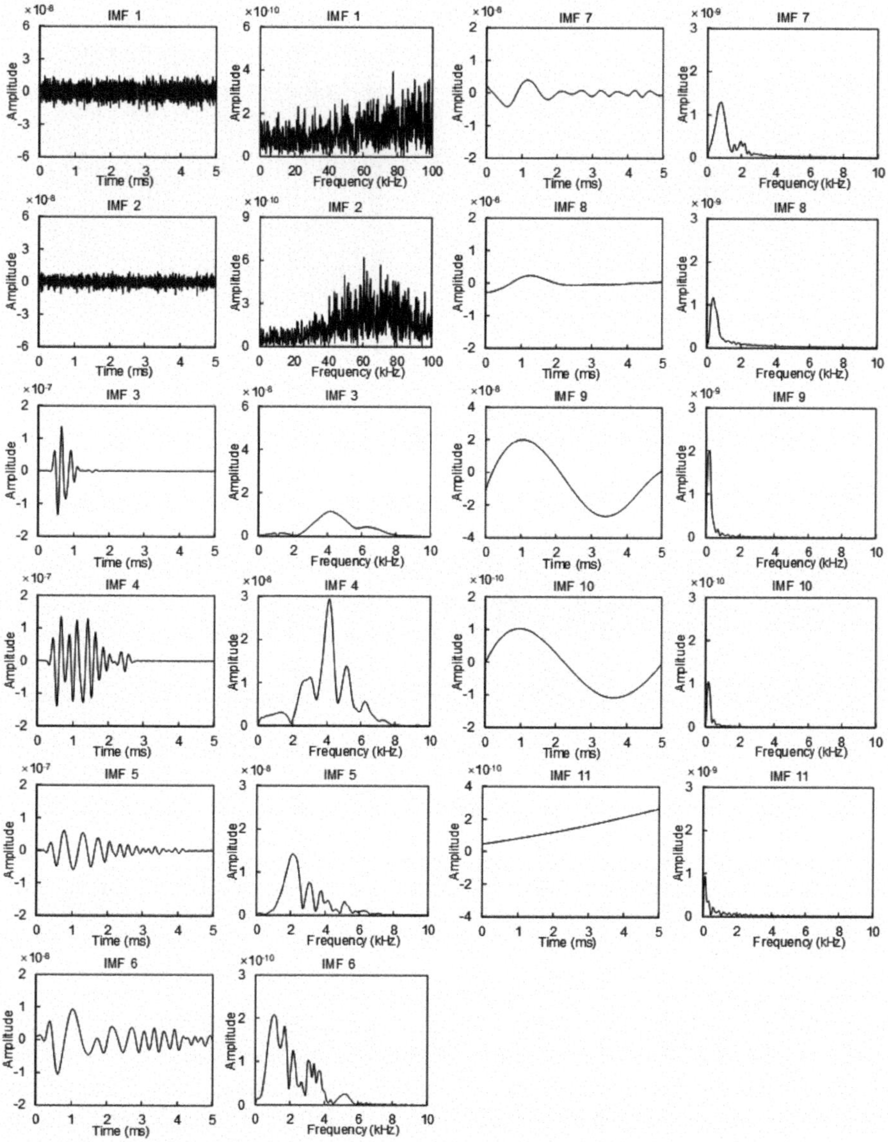


**Fig. 9.4** EEMD decomposed IMFs of No. 1 receiver model signal of Fig. 9.2a

receiver No. 2 by half a period throughout most of the recording. Although it may be challenging to visually discern, the two signals are approximately in phase at around 2 and 4 ms.

Figure 9.5c displays the absolute value of the instantaneous phase difference, while Fig. 9.5d showcases the corresponding IPDI. A time period (T) of 0.240 ms is utilized in Eq. 9.4 for IPDI calculation. Notably, two prominent peaks are observed

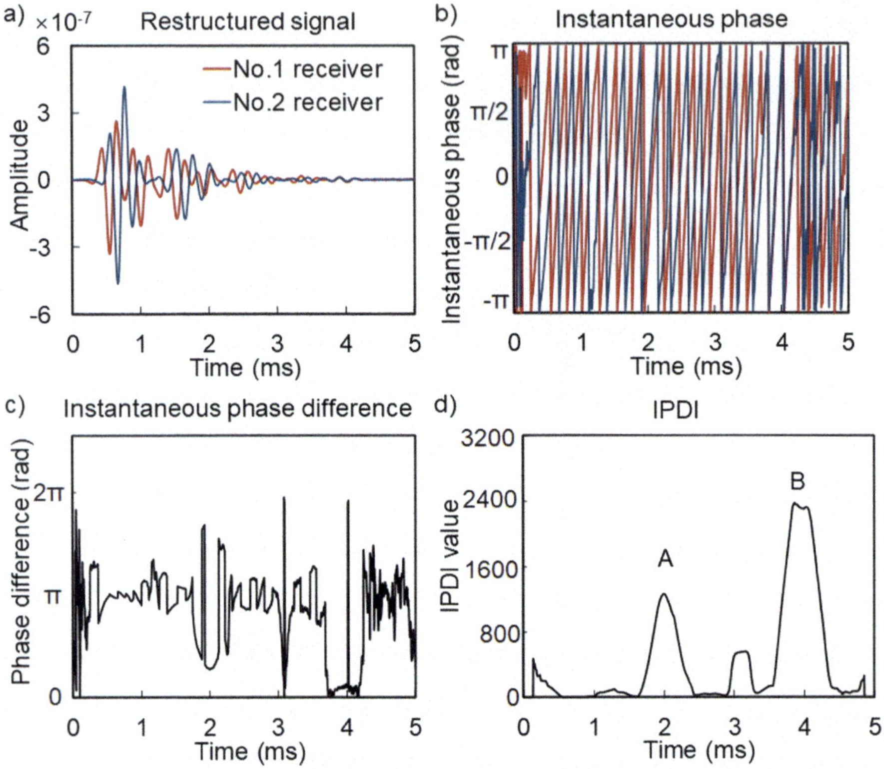


**Fig. 9.5** Instantaneous phase analysis of simulated waveforms shown in Fig. 9.2a. **a** The restructured signals obtained from the recordings of the two receivers are shown in Fig. 9.6a. **b** Fig. 9.6b displays the instantaneous phase of the restructured signals. **c** The absolute value of the instantaneous phase difference between the two restructured signals is depicted in Fig. 9.6c. **d** The IPDI, calculated using Eq. 9.4, is presented in Fig. 9.6d. Points A and B in Fig. 9.6d correspond to the reflections from the top and base of the cavity, respectively

in the IPDI plot (Fig. 9.5d). Peak A (1.986 ms) corresponds to the reflected PP-wave from the roof of the cavity, while Point B (3.894 ms) corresponds to the reflected PP-wave from the cavity floor. The arrival time of the direct wave is 0.320 ms. Table 9.1 provides the calculated arrival times of the reflected waves using Eq. 5. By employing the P-wave velocity values of 4000 m/s for bedrock and 2000 m/s for the water-saturated, clay-filled cavity, the distance of the karst cavity from the pile hole and its diameter can be estimated. It is worth noting that the identified errors are all below 5%, indicating the accuracy of the method. The IPDI value increases when the instantaneous phase of the two signals is more similar, enabling the identification of weak reflections from the cavity amidst the presence of interfering high-energy surface waves.

Subsequently, we employ the IPDI analysis method to investigate the impact of different scenarios outlined in Table 9.2. These scenarios include variations in the

**Table 9.1** Identified results of reflected wave

| Arrival time of reflected PP-wave from cavity roof (ms) | Arrival time of reflected PP-wave from cavity base (ms) | Calculated distance of karst cavity top from pile hole (m) | Calculated diameter of karst cavity (m) |
|---|---|---|---|
| 1.866 | 3.774 | 3.092 | 1.908 |

**Table 9.2** Simulated model parameters

| Scenario | Distance of karst cavity top from pile hole *Hc* (m) | Diameter of cavity *Dc* (m) | Coordinate of cavity center (x, y) (m) | Diameter of pile hole (m) | Material parameters of bed rock | Material parameters of slurry of pile hole | Material parameters of cavity infill |
|---|---|---|---|---|---|---|---|
| 1 | – | – | – | 1.6 | $v_p =$ 4000 m/s $v_s =$ 2300 m/s $\varepsilon_L =$ 10 $s^{-1}$ $\varepsilon_T =$ 50 $s^{-1}$ $\rho =$ 2300 kg/$m^3$ | $v_p =$ 2000 m/s $v_s = 0$ $\varepsilon_L =$ 1000 $s^{-1}$ $\varepsilon_T =$ 1500 $s^{-1}$ $\rho =$ 1500 kg/$m^3$ | – |
| 2 | 1 | 2 | (0, 2) | 1.6 | | | $v_p =$ 2000 m/s $v_s =$ 1100 m/s $\varepsilon_L =$ 2000$s^{-1}$ $\varepsilon_T =$ 3000 $s^{-1}$ $\rho =$ 1200 kg/$m^3$ |
| 3 | 2 | 2 | (0, 3) | 1.6 | | | |
| 4 | 3 | 2 | (0, 4) | 1.6 | | | |
| 5 | 4 | 2 | (0, 5) | 1.6 | | | |
| 6 | 3 | 2 | (0.5, 4) | 1.6 | | | |
| 7 | 3 | 2 | (1, 4) | 1.6 | | | |
| 8 | 3 | 2 | (1.5, 4) | 1.6 | | | |
| 9 | 3 | 2 | (2, 4) | 1.6 | | | |
| 10 | 3 | 0.5 | (0, 3.25) | 1.6 | | | |
| 11 | 3 | 1 | (0, 3.5) | 1.6 | | | |
| 12 | 3 | 1.5 | (0, 3.75) | 1.6 | | | |
| 13 | 3 | 2 | (0, 4) | 1.2 | | | |
| 14 | 3 | 2 | (0, 4) | 2.0 | | | |
| 15 | 3 | 2 | (0, 4) | 1.6 | | | $v_p =$ 1500 m/s $v_s = 0$ $\varepsilon_L =$ 1000 $s^{-1}$ $\varepsilon_T =$ 1500 $s^{-1}$ $\rho =$ 1000 kg/$m^3$ |
| 16 | 3 | 2 | (0, 4) | 1.6 | | | $v_p =$ 340 m/s $v_s = 0$ $\varepsilon_L =$ 5000 $s^{-1}$ $\varepsilon_T =$ 7000 $s^{-1}$ $\rho =$ 7.9 kg/$m^3$ |

distance between the cavity and the pile hole, the horizontal position of the cavity, the cavity diameter, the size of the pile hole, the cavity infill, and the spacing between the two receivers. By examining these different scenarios, we aim to evaluate their effects on the IPDI results.

(1) **Varying Cavity Distances from the Pile Hole**

Four different models were considered, each with a varying distance between the karst cavity top and the pile hole: 1, 2, 3, and 4 m. These models correspond to scenarios 2, 3, 4, and 5, respectively. Additionally, there is a scenario 1 without a cavity. The recorded signals from the No. 1 and No. 2 receivers are displayed in Fig. 9.6a and b, respectively. Through the utilization of the IPDI analysis method, the reflected PP-waves originating from the top of the cavities were detected at specific points: A1 (1.020 ms), A2 (1.418 ms), A3 (1.986 ms), and A4 (2.372 ms), as shown in Fig. 9.6c. Similarly, the reflected PP-waves from the base of the cavities corresponded to points B1 (2.922 ms), B2 (3.358 ms), B3 (3.894 ms), and B4 (4.368 ms). Comparatively, the IPDIs obtained from the signals without a cavity (represented by the solid line) exhibited significantly lower amplitudes. The arrival time of the direct waves was determined as 0.320 ms from the original modeling signal. Equivalently, the arrival times for the cavity top and base were calculated using Eq. 5. Based on the observed arrival times and known material velocities, the distance of the karst cavity top and base from the pile hole, as well as the diameter of the karst, was estimated. A summary of the results can be found in Table 9.3.

In Fig. 9.6c, it can be observed that the IPDI reflections from both the top and base of the cavity exhibit an increasing trend as the distance of the cavity from the pile hole increases. This is attributed to the decrease in instantaneous phase difference between the two received signals as the reflected wave travels a longer path.

(2) **Varying Cavity Horizontal Locations**

For scenarios 4, 6, 7, 8, and 9, five karst cavities were modeled with a fixed depth of 3 m and diameter of 2 m. However, the horizontal locations vary from 0 to 0.5, 1.0, 1.5, and 2.0 m. The test signals recorded by the No. 1 and No. 2 receivers are presented in Fig.9.7a and b. The arrival of reflected PP-waves from the top of the cavities at the receivers occurs around 1.9 ms, making it visually challenging to distinguish the difference in arrival times. By employing IPDI calculations, the reflected PP-waves from the top of the cavities were identified at points A1 (1.986 ms), A2 (2.008 ms), A3 (1.932 ms), A4 (2.140 ms), and A5 (2.170 ms) in Fig. 9.7c. Similarly, the reflected PP-waves from the base of the cavities corresponded to points B1 (3.894 ms), B2 (3.920 ms), B3 (3.906 ms), B4 (4.136 ms), and B5 (4.074 ms). In Fig. 9.7c, it is worth noting that there is an IPDI peak (point C2) around 3 ms, which corresponds to the weak reflected PS-waves from the top of the cavity. This occurs because the P-wave receiver is capable of detecting the weak vertical component of reflected PS-waves as well. To estimate the horizontal and vertical coordinates of the cavities, two sets of signals at different azimuths are required.

Taking scenario 7 as an example, where the cavity center is offset by 1 m from the center of the pile hole. In Fig. 9.8, the IPDI is computed using two sets of test signals acquired from the instrument array, with one array aligned as shown in Fig. 9.1 and the other array rotated by 180 degrees. The reflected PP-waves from the top of the cavity are identified as point A1 (1.932 ms) and A2 (2.042 ms). Consequently, utilizing Eq. 9.5, the corresponding arrival times are calculated as 1.812 and 1.922 ms.

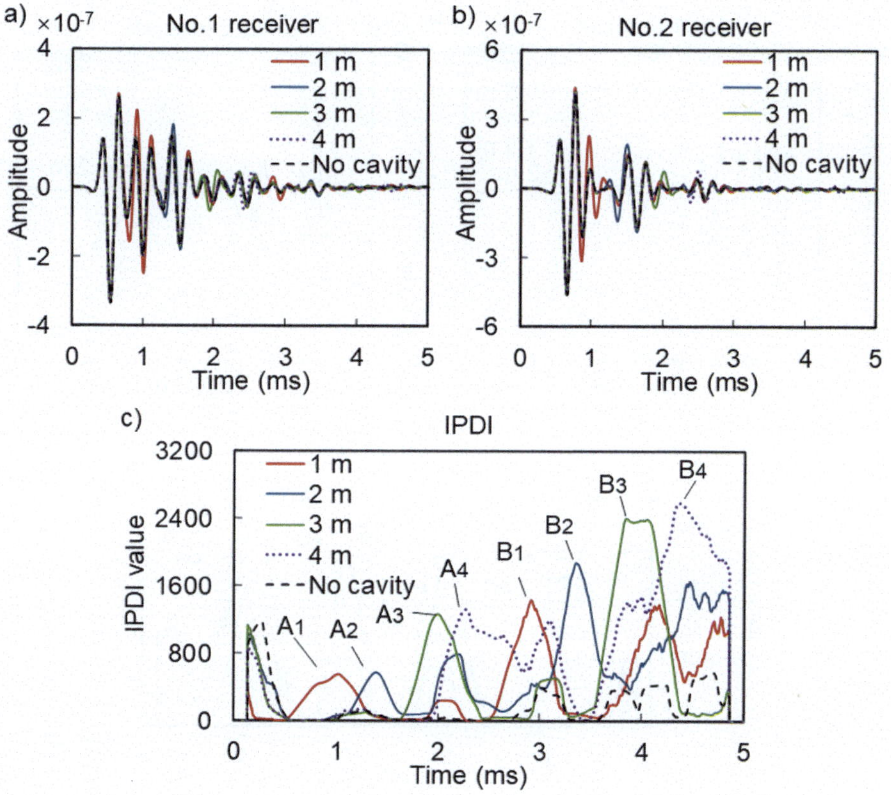


**Fig. 9.6** Modeling results of scenarios 1, 2, 3, 4, and 5. **a** Signals recorded by No. 1 receiver. **b** Signals received by No. 2 receiver. **c** Corresponding IPDI analysis

**Table 9.3** Estimated distance and diameter of the cavity

| Distance of karst cavity top from pile hole (m) | Arrival time of reflected PP-wave from cavity roof (ms) | Arrival time of reflected PP-wave from cavity floor (ms) | Calculated distance of karst cavity top from pile hole (m) | Calculated diameter of cavity (m) |
|---|---|---|---|---|
| 1 | 0.900 | 2.802 | 1.160 | 1.902 |
| 2 | 1.298 | 3.238 | 1.956 | 1.940 |
| 3 | 1.866 | 3.774 | 3.092 | 1.908 |
| 4 | 2.252 | 4.248 | 3.864 | 1.996 |

To determine the location of the karst cavity top, the arrival times of the reflected PP-waves and direct waves (0.32 ms) are employed in the following equation:

$$(x - x_n)^2 + (y - y_n)^2 = \left((t_{An} - t_{Bn}) \times v_p/2\right)^2. \tag{9.6}$$

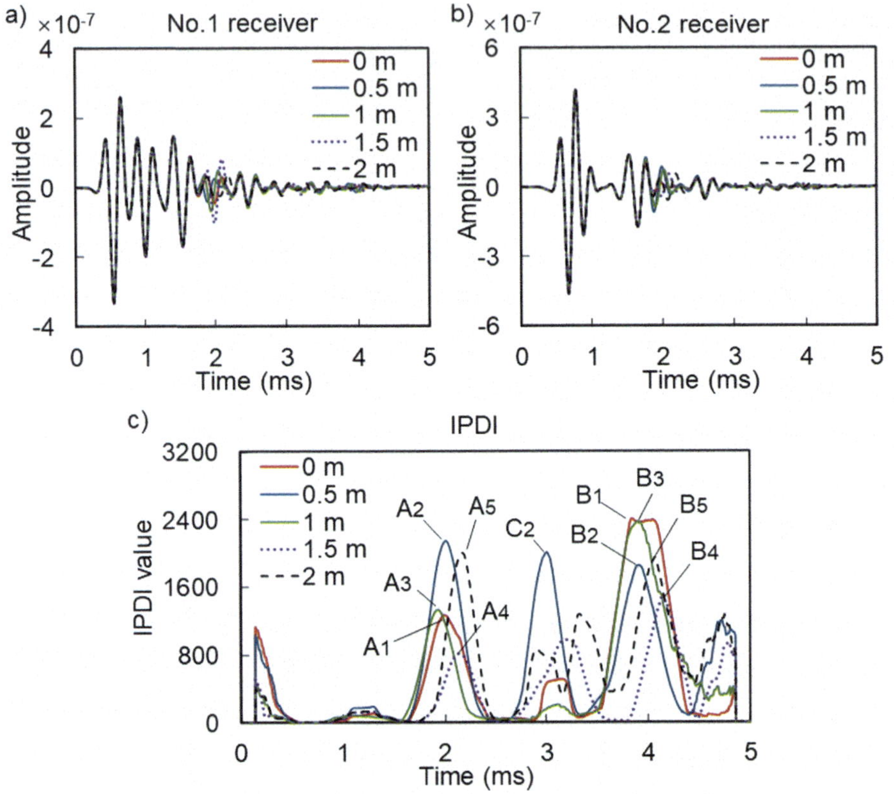


Fig. 9.7 Modeling results of scenarios 4, 6, 7, 8, and 9. **a** Signals recorded by No. 1 receiver. **b** Signals received by No. 2 receiver. **c** Corresponding IPDI analysis

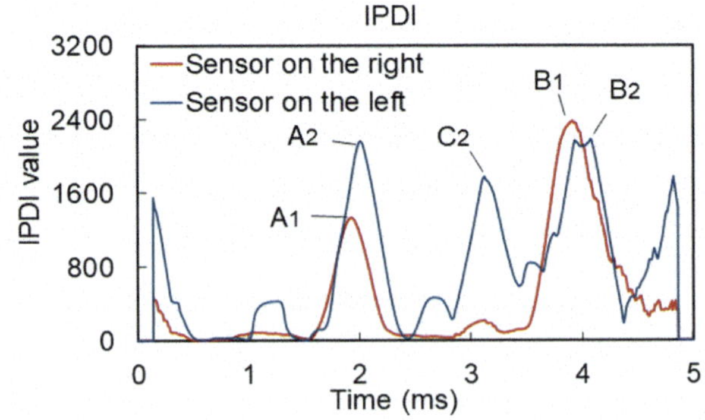


**Fig. 9.8** IPDI calculated by two pairs of modeled signals. The coordinates of the right pair of receivers are (0.4, 0 m) and (0.6, 0 m). The coordinates of the left pair of receivers are (−0.4, 0 m) and (−0.6, 0 m)

The coordinates of the reflector on the top of the cavity are denoted as (x, y), while $(x_n, y_n)$ (where $n = 1, 2$) represents the coordinates of No. 1 receivers. The arrival time of the cavity top PP-reflected waves and direct waves is denoted as $t_{An}$

and $t_{Bn}$, respectively (where $n = 1, 2$). By solving the equation, the solution yields $x = 0.729$ m and $y = 2.998$ m. Consequently, the coordinates of the reflection point on the top of the cavity are (0.729, 2.998 m), which is in agreement with the top of the cavity model (1, 3 m).

(3) **Varying Cavity Diameters (Size)**

To examine the capability of determining the size of karst cavities using the IPDI method, scenarios 4, 10, 11, and 12 were modeled. The diameters of the karst cavities were varied, ranging from 0.5 m to 1.0 m, 1.5 m, and 2.0 m, which also affected the center positions of the cavities. The modeled waveforms are depicted in Fig. 9.9a and b. Figure 9.9c demonstrates that the reflected PP-waves from the top of the cavities can be identified at specific points: A1 (1.946 ms), A2 (1.904 ms), A3 (1.914 ms), and A4 (1.986 ms). Correspondingly, the reflected PP-waves from the base of the cavities correspond to different points, including B1 (2.702 ms), B2 (2.954 ms), B3 (3.368 ms), and B4 (3.894 ms). Consequently, the estimated diameters of the karst cavities are as follows: 0.756, 1.050, 1.454, and 1.908 m, respectively. It is noteworthy that the accuracy of the estimated diameters improves as the size of the cavity increases.

(4) **Pile Hole and Instrumentation Geometry Effects**

The recorded signals can be influenced by various factors such as the diameter of the pile hole, the spacing between receivers and the transmitter, as well as the cavity infill. In order to explore the impacts of these factors on the IPDI analysis method, scenarios 4, 13, 14, 15, and 16 were conducted. The outcomes of these scenarios are presented in Fig. 9.10.

To investigate the influence of pile hole diameter on the IPDI analysis method, scenarios 4, 13, and 14 were conducted while maintaining the same positions for the two receivers. In Fig. 9.10a, points A2 (1.986 ms), A3 (2.028 ms), B2 (3.894 ms), and B3 (3.920 ms) accurately identify the cavity reflection. However, point A1 (1.798 ms) deviates from the other two estimates due to the placement of the No. 2 receiver in scenario 13, which is located at the corner of the hole wall and the hole bottom. The recorded signal at the corner contains complex conversions of wave types, leading to inaccurate calculation of the instantaneous phase. Therefore, it is recommended to position the receivers in the center area of the pile hole bottom, away from the pile hole walls. This ensures more reliable and precise calculations in the IPDI analysis method Three scenarios (4, 15, and 16) with different cavity infill materials were used to investigate the IPDI method.

To simulate various cavity infill materials, different impedances of P-waves ($\varepsilon_L$) and S-waves ($\varepsilon_T$), as well as wave velocities, are utilized (refer to Table 9.2). In Fig. 9.10b, the calculated IPDIs reveal that the reflected P-waves from the top of the cavity correspond to 1.946 ms (point A1 in Fig. 9.12b), 1.934 ms (point A2), and 1.986 ms (point A3). Similarly, the reflected P-waves from the cavity base correspond to 4.402 ms (point B1 in Fig. 9.10b) and 3.894 ms (point B3). Based on these calculations, the diameters of the water-filled and clay-filled cavities are estimated to be 1.842 and 1.908 m, respectively. However, no visible IPDI peak is observed

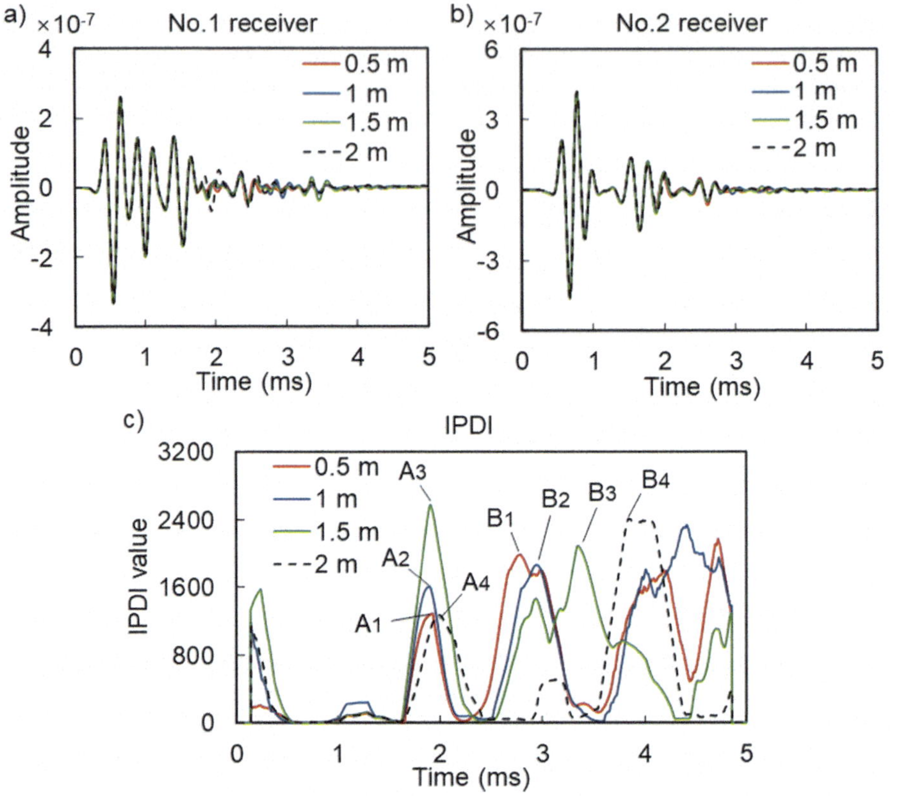


**Fig. 9.9** Modeling results of varying cavity diameters (scenarios 4, 10, 11, and 12). **a** Signals recorded by No. 1 receiver. **b** Signals received by No. 2 receiver. **c** Corresponding IPDI analysis

corresponding to the reflection from the air-filled cavity base. This is due to the slow velocity of propagation through air and the insufficient recording time, which does not capture the reflection from the cavity base.

To investigate the impact of the separation between the transmitter and receiver, three pairs of receivers in scenario 4 were selected for analysis. The pairs of receivers had different separations: OA = 0.3 m and OB = 0.5 m, OA = 0.4 m and OB = 0.6 m, and OA = 0.5 m and OB = 0.7 m, as depicted in Fig. 9.1. It is worth noting that the spacing of 0.2 m between the two receivers in each scenario remained constant. As demonstrated in Fig. 9.10c, the reflected P-waves from the top and base of the cavities were correctly identified at points A1 (1.962 ms), A2 (1.986 ms), and A3 (1.998 ms), and points B1 (4.060 ms), B2 (3.894 ms), and B3 (4.088 ms), respectively. An IPDI peak (point C3) around 3 ms was observed, which was attributed to the vertical component of the reflected PS-wave from the top of the cavity. This phenomenon is similar to what was observed in scenario 6 (Fig. 9.7). It suggests that the vertical component of reflected PS-waves has the potential to interfere with cavity detection in the IPDI method.

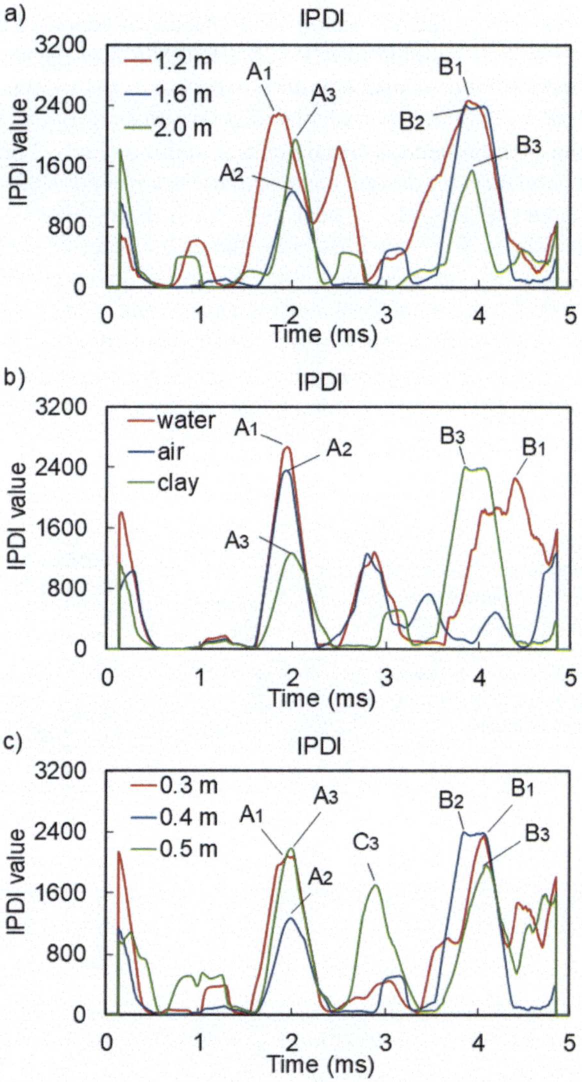


**Fig. 9.10** **a** IPDIs of the scenarios (4, 13, and 14) with different diameter pile holes. **b** IPDIs of the scenarios (4, 15, and 16) with different cavity infills. **c** IPDIs of scenario 4 with different spacing between No. 1 receiver and the transmitter

### (5) **Receiver Spacing Effects**

In order to distinguish the reflected P-wave from surface waves, it is recommended to have a desirable spacing between two receivers of approximately 0.5 times the wavelength ($\lambda$) of surface waves. For example, in Fig. 9.1, a spacing of AB = 0.2 m was chosen, considering that the wavelength is approximately 0.4 m. However, in practical situations, the spacing between receivers may not precisely match half the wavelength, especially when estimating the surface-wave velocity at the bottom of the pile hole based on lithology identification during drilling. To assess the sensitivity of receiver spacing and its impact on the effectiveness of the method, various geophone separations ranging from 0.125 λ to 0.875 λ of the surface wave in scenario 4 were analyzed. For instance, with a fixed OA = 0.4 m, the spacing AB was varied from 0.05 to 0.35 m. The calculated IPDI values in Fig. 9.11a and b accurately identify the cavity reflection (IPDI peaks at approximately 2 ms), but with differing levels of confidence. By design, the baseline (background) IPDI value without void reflections should be close to or near zero when the receivers are set at a spacing of 0.5 λ. However, Fig. 9.11 illustrates that as the receiver separation deviates from the 0.5 λ spacing, the baseline values progressively increase, making it more challenging to differentiate the peaks of the cavity reflection from noise peaks. Therefore, even if an accurate bedrock surface-wave velocity is not initially available, in practical scenarios, the geophone interval can be adjusted based on a rapid on-site calculation of the IPDI using recorded data. The geophone spacing is considered correct when the background IPDI values are close to zero.

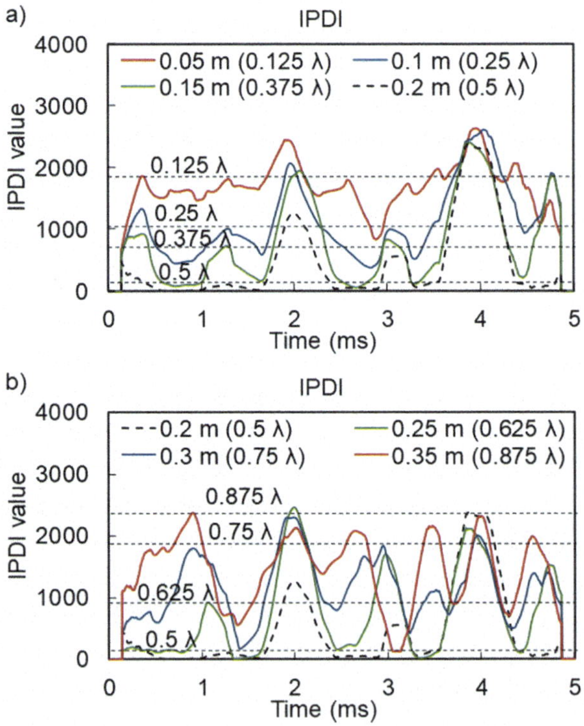


**Fig. 9.11** IPDIs computed for receiver spacing shorter and longer than the optimum 0.5 wavelength (λ). **a** Spacing of 0.125, 0.25, 0.375, and 0.5 λ, respectively. **b** Spacing of 0.5, 0.625, 0.75, and 0.875 λ, respectively

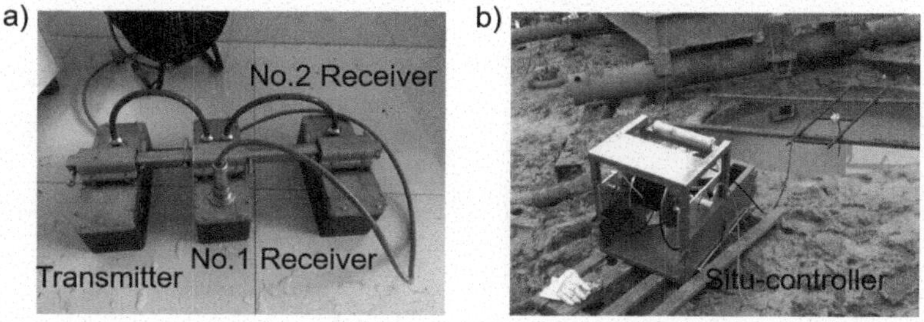


**Fig. 9.12** **a** Downhole sonar consisting of a transmitter and two receivers. **b** Surface controller connected to the downhole tool using cable

## 9.5 Field Investigation—Site 1

A field test was conducted using a sonar detection prototype, which is illustrated in Fig. 9.12. The prototype comprises a newly developed downhole tool equipped with two receivers and a surface controller. The separation between the transmitter and the two receivers can be customized by adjusting the positions of the components along the connecting rod.

The field test was carried out at a commercial industry park under construction in Guangxi province, located in South China. Prior to the test, a comprehensive geological investigation of the construction site was conducted. The investigation revealed the presence of various layers, including an approximately 2-m thick Quaternary topsoil, followed by a 2.5-m thick brecciated Quaternary Silty clay layer, and finally, a medium-weathered brecciated limestone layer exceeding 5 m in thickness. The specific bored pile that was tested, referred to as No. 43–2 and depicted in Fig. 9.12b, had a depth of 8.7 m and a diameter of 1.6 m. The sonar detection prototype was tested before the installation of the steel reinforcement cage. For the test, the downhole tool was positioned at the center of the pile hole bottom with the assistance of a surface ground support system, as shown in Fig. 9.12b. The separation between the transmitter and the first receiver was set at 0.3 m, while the separation between the two receivers was 0.25 m. The recorded signals from the two receivers are presented in Fig. 9.13a.

After employing the Ensemble Empirical Mode Decomposition (EEMD) technique on the signals, reconstructed signals are obtained by summing up the Intrinsic Mode Functions (IMFs) with dominant frequencies exceeding 1 kHz. These reconstructed signals are presented in Fig. 9.13b, and they exhibit characteristics of stationarity and narrow bandwidth. The proposed method includes instantaneous phase analysis, which is showcased in Fig. 9.13c and d. The Instantaneous Phase Difference Image (IPDI) reveals a significant peak (point A at 2.160 ms), indicating a possible PP-reflected wave originating from the roof of a cavity. By considering the dominant frequency of the test signal (2.929 kHz) and utilizing Eq. 9.5, the arrival time of the PP-reflected wave is calculated as 1.989 ms. Additionally, the arrival time of the direct wave is determined to be 0.140 ms from the original test signal. Considering

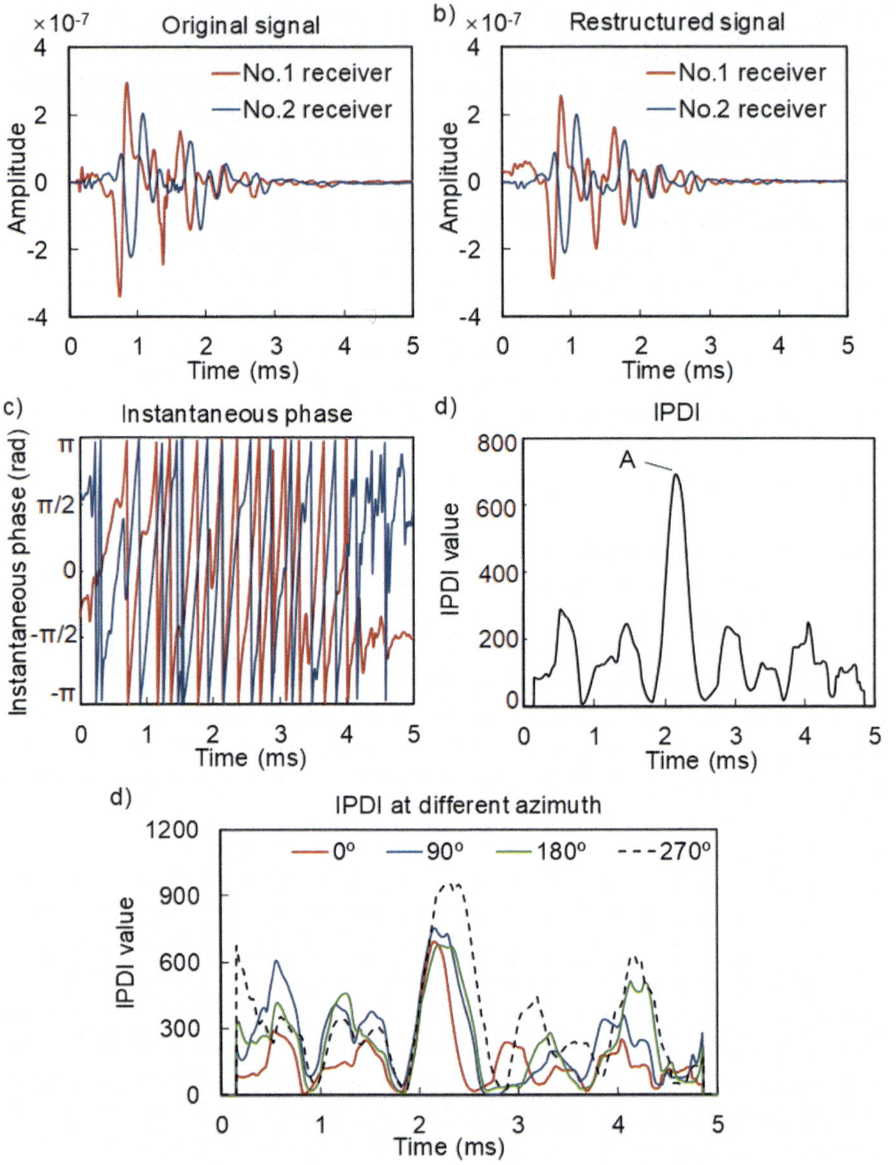


**Fig. 9.13** Instantaneous phase analysis of two test signals at bored pile (43–2). **a** Original test signals of two receivers. **b** Corresponding reconstructed signals. **c** Instantaneous phase of reconstructed signals. **d** Instantaneous phase difference intensities calculated by Eq. 9.4. **e** Calculated IPDI of four azimuthal detection test signals

the P-wave velocity in the weathered limestone breccia to be approximately 3100 m/s, the depth of the potential cavity roof can be estimated to be 2.9 m.

Test signals were collected using the downhole probe, with four pairs obtained at different azimuths: 0°, 10°, 180°, and 270°. The IPDIs of these signals are displayed in Fig. 9.13e. Remarkably, all four recordings exhibit nearly identical arrival times of the PP-reflected waves, further confirming the efficacy of the proposed method. This similarity in arrival times indicates that the cavity is likely located directly beneath the pile hole. However, in cases where the cavity is situated off-center, a notable difference in the arrival times can serve as an indication of the azimuth or direction of the potential cavity. By analyzing and comparing the arrival times from different azimuths, valuable insights can be gained regarding the location and orientation of the cavity.

The IPDI detection analysis results were verified using advanced drilling borehole optical imaging. The image of the borehole wall, depicted in Fig. 9.14, confirms the presence of a cavity starting at a depth of 11.2 m, with the distance between the pile bottom and cavity roof measuring approximately 2.5 m. It is worth noting that there is a slight difference between the estimated depth from IPDI (around 2.9 m) and the borehole imaging (approximately 2.5 m). This variation could be attributed to the lower P-wave velocity of the bedrock in the proximity of the cavity. The borehole image also reveals that the cavity is partially filled with clay, while the upper portion remains air-filled. Consequently, the reflected P-wave from the floor of the cavity cannot be identified in this particular case. Based on the detection results, it is recommended to deepen the pile hole. Additionally, steel casing is necessary within the depth range of the cavity to prevent the drilling pile hole from collapsing.

## 9.6 Field Investigation—Site 2

At a bridge foundation project in Guangdong, South China, another field sonar detection was carried out. Prior to the construction of the bridge piles and testing of the pile hole sonar technology, conventional cross-hole seismic tomography was employed to investigate the subsurface. Two cross-hole seismic tomography profiles were conducted, spanning across three boreholes. Figure 9.15a illustrates the velocity tomography results alongside the borehole logs. Within the tomograms, multiple karst voids were identified as low-velocity anomalies. During the drilling of a pile hole at the location of borehole Y64-2, the sonar detection system was deployed twice. The first deployment took place when the bottom of the hole reached a depth of 34.9 m, and the second deployment occurred at a depth of 48.7 m. The diameter of the pile hole was measured as 1.8 m, and the downhole detector was positioned at the center of the pile hole. For the sonar detection, the separation between the transmitter and the first receiver was set at 0.35 m. After several trials, the separation between the two receivers was determined as 0.18 m, using the criterion of achieving a low background IPDI value.

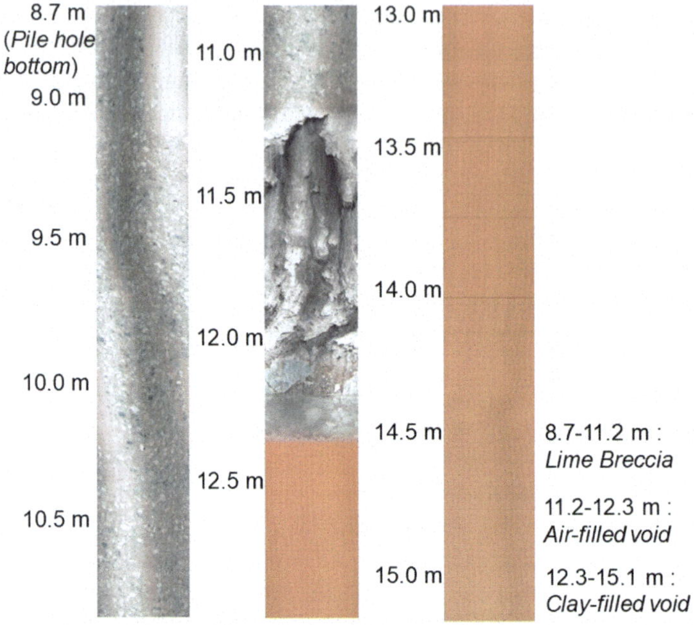


**Fig. 9.14** Advanced drilling borehole optical image beneath pile hole

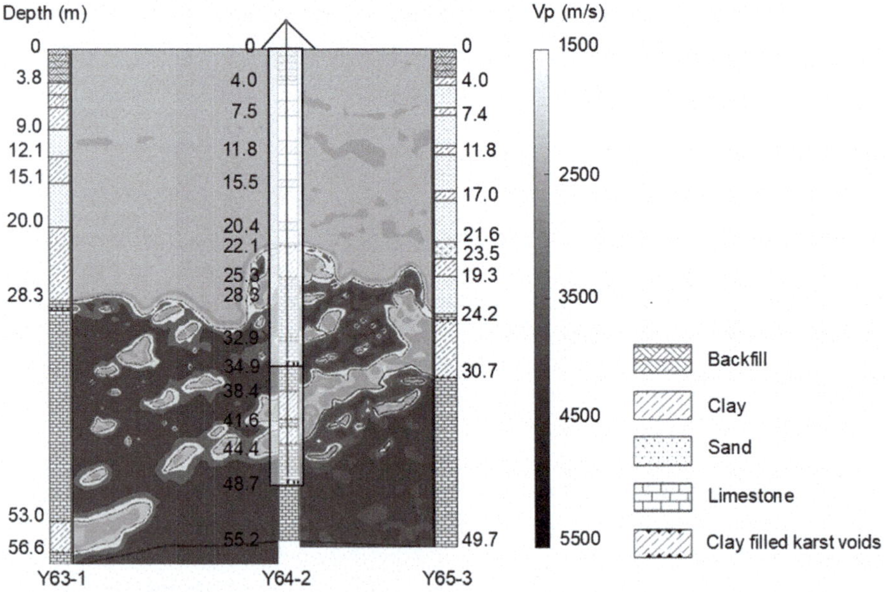


**Fig. 9.15** Cross-hole seismic tomography between boreholes Y63-, Y64-2, and Y65-3 along with borehole logs identifying subsurface lithologies and voids

Figure 9.16a and c displays the recorded signals from the two receivers at different depths. After applying the Ensemble Empirical Mode Decomposition (EEMD) and selecting specific Intrinsic Mode Functions (IMFs), the corresponding reconstructed signals are obtained, as shown in Fig. 9.16b and d. The analysis of the Instantaneous Phase Difference Image (IPDI) for the detection signals at the depth of 34.9 m, presented in Fig. 9.16e, exhibits a significant peak (2.420 ms), indicating a potential PP-reflected wave. However, no notable peak is observed above the background noise in the IPDI of the signals at the depth of 48.7 m. Based on Eq. 9.5, the arrival time of the PP-reflected wave can be calculated as 2.285 ms, with a dominant frequency of 3.711 kHz. Additionally, the arrival time of the direct wave is identified as 0.230 ms from the original test signal. Considering the P-wave velocity in the bedrock to be approximately 3600 m/s, as determined by the cross-hole tomography, the detection at the depth of 34.9 m suggests a potential cavity roof located 3.7 m below the pile hole, at a depth of 38.6 m. Conversely, the deeper detection at the depth of 48.7 m indicates that the bedrock below the pile hole remains intact. These two sonar detection results align with the cross-hole tomography findings and the initial Y64-2 borehole log.

## 9.7 Further Discussion

Numerical simulations and field tests have indicated the effectiveness of the IPDI-based method in detecting voids by distinguishing the weak reflected PP-wave of a cavity from multiple reflected surface waves originating from the wall of the pile hole. Several aspects of the method warrant further discussion, including the use of Ensemble Empirical Mode Decomposition (EEMD) for signal processing.

Use of EEMD: EEMD is employed to filter out frequency noise and reconstruct signals while preserving accurate phase information. Field test signals are prone to contamination from various sources such as noise in the field environment, electronic circuitry in the instrument, and vibrations from the transmitter and receivers. Effective frequency filtering is crucial in signal analysis. Furthermore, instantaneous phase serves as a vital indicator in the IPDI-based method. However, designing a traditional bandpass frequency filter, such as FIR and IIR filters, that can retain complete and accurate phase information is challenging. EEMD offers an adaptive and envelope-based decomposition and reconstruction method, addressing the mode mixing phenomenon found in Empirical Mode Decomposition (EMD). By adding white noise to the original signal, EEMD achieves better resolution in frequency filtering, resulting in improved preservation of accurate phase information.

Identification of Surface-Wave Wavelength: Determining the wavelength of surface waves is a crucial step in the IPDI-based method. It is recommended to set the interval between two receivers as half the wavelength of the multiple reflected surface waves originating from the wall of the pile hole. This configuration leads to a more pronounced IPDI peak and background values that are close to zero when appropriately chosen. The wavelength can be estimated by considering the dominant

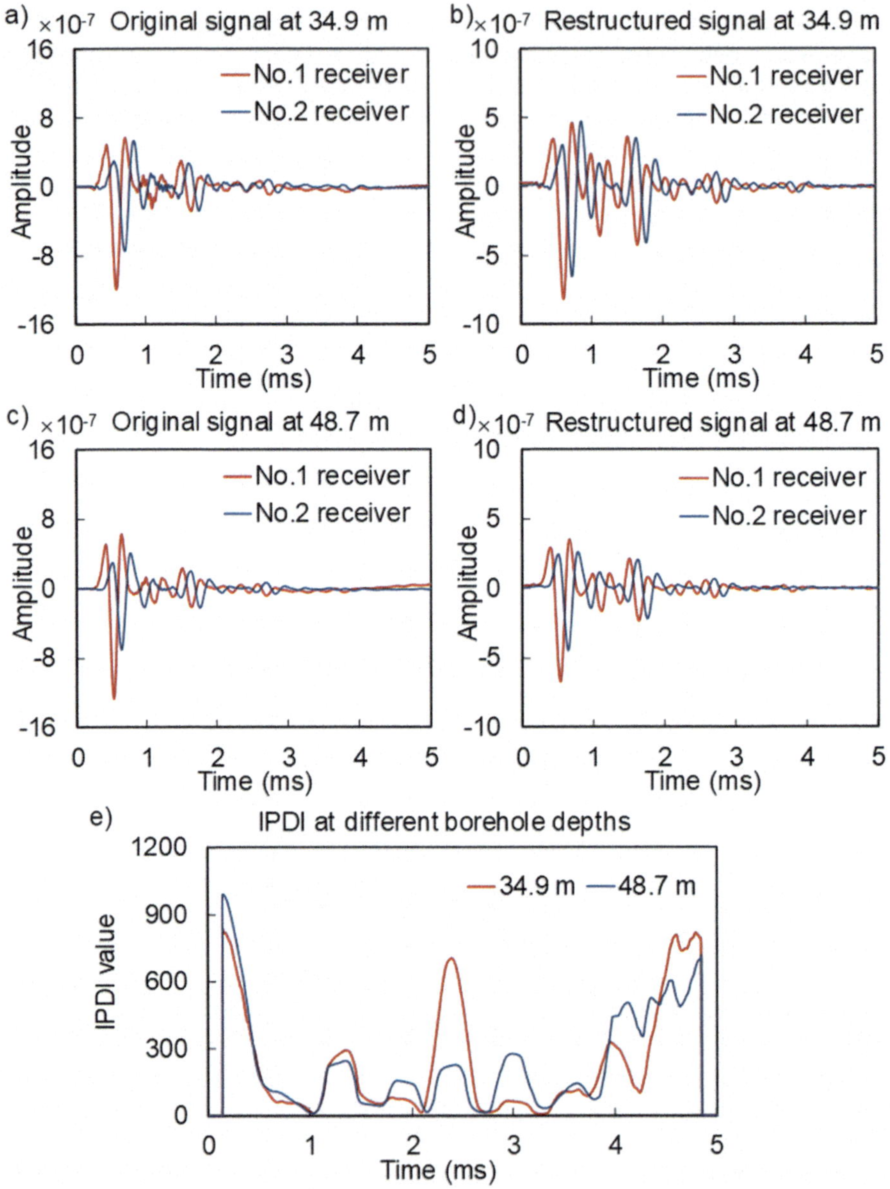


**Fig. 9.16** Instantaneous phase analysis of test signals at two different depths of pile (Y64-2). **a** Original test signals of two receivers at depth of 34.9 m. **b** Reconstructed signals corresponding to the signals in (**a**). **c** Original test signals of two receivers at depth of 48.7 m. **d** Reconstructed signals corresponding to the signals in (**c**). **e** Calculated IPDI of two detection test signals at different depths

frequency of the test signal and the velocity of the surface waves. The dominant frequency can be obtained through spectrum analysis of the recorded signals. The velocity of surface waves is typically around 0.9 times the shear-wave velocity of the bedrock, which can be determined using geotechnical data specific to the site.

IPDI Threshold: The IPDI value significantly increases when a cavity is present beneath the pile. The IPDI associated with the cavity reflection can be up to 5 times larger than the maximum IPDI observed in models without a cavity. However, establishing a universal threshold for IPDI is challenging due to the variability of geological environments, pile conditions, and void conditions. For instance, the threshold of IPDI is lower when the cavity is located in close proximity to the pile hole. Moreover, Eq. 9.4 indicates that the IPDI threshold also depends on the dominant frequency and sampling rate of the signals. Consequently, calibrating the IPDI threshold is an ongoing research topic. Efforts are being made to develop a more comprehensive understanding of the factors influencing the IPDI threshold and to establish calibration procedures that account for the specific conditions of each scenario.

Sensitivity of the IPDI Method: A comprehensive sensitivity study was performed to assess the impact of material and instrumentation geometry parameters using synthetic modeling in 16 different scenarios. The IPDI-based method proved effective in determining the depth, horizontal location, and size of karst cavities. However, among the parameters examined, geophone spacing had a greater influence on the sensitivity of the IPDI method compared to the interval between the transmitter and receiver. When the geophone separation is set close to half of the wavelength, the IPDI method exhibits significantly higher IPDI peaks compared to background values and noise. In practical applications, multiple trials can be conducted by adjusting the geophone spacing along the connecting rod until the background IPDI value approaches zero. This iterative process helps optimize the geophone spacing for enhanced sensitivity and accurate detection of voids. Another critical parameter is the integration range in Eq. 9.4. Adjusting the integration range can mitigate plateau-shaped IPDI peaks, as observed in point B in Fig. 9.5d. Taking scenario 4 as an example, increasing the integration range enhances the clarity of point B in Fig. 9.17 but may result in plateau-shaped IPDI peaks at point C. A default integration range of one period is commonly used, as it often yields satisfactory results.

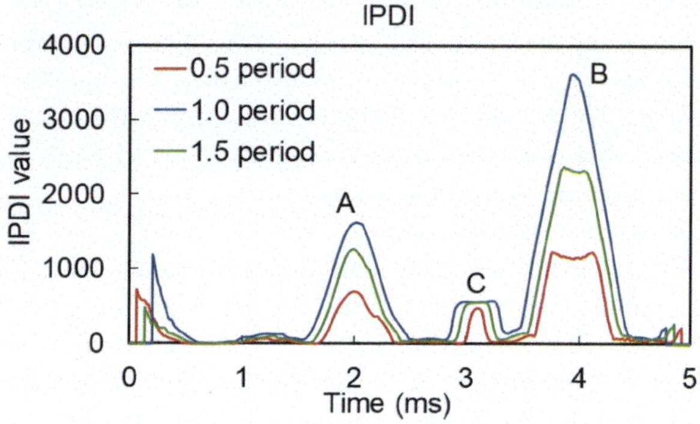


**Fig. 9.17** IPDIs of the scenario 4 with varying the integration ranges in Eq. 9.4

Limitations of the IPDI Method: Despite the success of the IPDI method in void detection, several limitations should be acknowledged. These limitations include:

Boundary Effect of EEMD: The accuracy of the IPDI method can be affected by artifacts caused by the boundary effect of Ensemble Empirical Mode Decomposition (EEMD). Some IPDI peaks observed at the beginning and end of the signal may be attributed to this effect (e.g., Fig. 9.16e). To mitigate this issue, it is recommended to extend the length of the recorded signal, which can help minimize the impact of boundary artifacts.

Ultra-shallow Karst Cavities: The IPDI method may not perform optimally for detecting ultra-shallow karst cavities located within 0.5 m beneath the pile hole. However, in such cases, the high amplitude of PP-reflections can often be identified directly from the original signal without the need for applying the IPDI method.

Azimuthal Detection: The identification of the azimuth of karst voids relies on the observation of different arrival times of the cavity reflection when the downhole tool is oriented at different azimuths within the pile hole. However, in certain instances, the difference in arrival times may be too minimal to differentiate accurately through azimuthal detection by rotating the downhole tool.

Poor Pile Hole Construction Quality: Factors such as the accumulation of sediment or the presence of inclined pile holes can negatively impact the detection accuracy. Sediment accumulation at the bottom of the pile hole introduces additional reflections and surface-wave dispersion, thus complicating the analysis. It is crucial to clean up any sediment present in the pile hole before conducting the detection. Additionally, an inclined pile hole makes it challenging to accurately position the transmitter at the center of the pile hole.

## 9.8 Summary

In this chapter, we introduce an advanced sonar method for detecting karst cavities beneath piles. The method is centered around the analysis of instantaneous phase differences between two receiver recordings, enabling the identification of reflected P-waves from a cavity. The key conclusions drawn from this study are as follows:

Differentiating weak cavity reflections from multiple reflected surface waves using conventional processing methods poses challenges. Our proposed method employs two receivers positioned at a distance equal to half the wavelength of surface waves. By design, the instantaneous phase difference of reflected P-waves from a cavity between the two receivers is close to zero, while the instantaneous phase difference of multiple reflected surface waves is $\pm\pi$. This allows for the extraction of weak cavity reflections.

We define a new index called Instantaneous Phase Difference Intensity (IPDI) that quantifies the similarity of instantaneous phase between signals at the two receivers. Higher IPDI values indicate that the two test signals exhibit similar instantaneous phase, indicating the arrival of a reflection from a cavity. This allows for the identification of the first break of reflected P-waves from the roof and floor of the cavity. The calculation of instantaneous phase differences in the recorded signals is performed using Hilbert Transform (HT) after mitigating frequency noise and reconstructing the signals using EEMD.

The IPDI-based signal analysis method is applied to numerical models, demonstrating the successful decomposition and reconstruction of signals while preserving accurate phase information through the use of EEMD. Despite strong contamination from multiple reflected waves and noise, the IPDI peak serves as an effective indicator for the arrival of reflections from a cavity. The method proves capable of determining the depth, horizontal location, and size of the karst cavity.

Variation in recorded signals can be attributed to factors such as pile hole diameter, cavity infill material, spacing between the transmitter and receivers, and spacing between the two receivers. However, numerical model analyses indicate that the IPDI-based method performs well even as these parameters vary. It is worth noting that the identification of the cavity base reflection may be affected by the presence of the vertical component of reflected PS-waves from the cavity.

A prototype based on the IPDI method was developed and deployed in two field studies. The validity of IPDI void detections was confirmed through direct drilling and borehole tomography. The separation between the two receivers can be adjusted manually based on the wavelength of multiple surface waves. Azimuthal detection is achieved by rotating the downhole probe and utilizing a 3D electric compass within the tool. The field study results demonstrate the effectiveness of the IPDI-based prototype in accurately locating karst cavities beneath piles.

## References


Battista BM, Knapp C, McGee T, Goebel V (2007) Application of the empirical mode decomposition and Hilbert-Huang transform to seismic reflection data. Geophysics 72(2):H29–H37

Craig MS, Genter RL (2006) Geophone array formation and semblance evaluation. Geophysics 71(1):Q1–Q8

Fomel S (2013) Seismic data decomposition into spectral components using regularized nonstationary auto regression. Geophysics 78(6):O69–O76

Honório B, De Matos M, Vidal A (2017) Progress on empirical mode decomposition-based techniques and its impacts on seismic attribute analysis. Interpretation 5:SC17–SC28

Huang NE, Shen Z, Long SR, Wu MC, Shi HH, Zheng Q (1998) The empirical mode decomposition and the Hilbert spectrum for nonlinear and non-stationary time series analysis. Proceed Royal Soc London A 454:903–995

Liu L, Shi Z, Peng M, Liu C, Tao F, Liu C (2018) Numerical modeling for karst cavity sonar detection beneath bored cast in situ pile using 3D staggered grid finite difference method. Tunnel Underground Space Technol 82:50–65

Shi Z, Liu L, Peng M, Tao F, Liu C (2016) A sonar detection technology for karst cavities under bored piles and its application. Chin J Rock Mech Eng 35:177–186

Shi Z, Liu L, Peng M, Liu C, Tao F, Liu C (2018) Sonar detection for karst cavities and fracture under bored pile by generalized S-transform: GeoShanghai International Conference. Springer, Singapore, pp 525–533

Taner MT, Koehler F, Sheriff RE (1979) Complex seismic trace analysis. Geophysics 44:1041–1063

Torres ME, Colominas MA, Schlotthauer G, Flandrin P (2011) A complete ensemble empirical mode decomposition with adaptive noise. In: IEEE annual international conference on acoustics, speech and signal processing (ICASSP)
Wu Z, Huang N (2009) Ensemble empirical mode decomposition: a noise-assisted data analysis method. Adv Adapt Data Anal 1:1–41
Wu CF, Huang HC (2019) Detection of a fracture zone using microtremor array measurement. Geophysics 84(1):B33–B40

# Chapter 10
# Complete Application

## 10.1 A Complete Numerical Validation

During the investigation phase, borehole sonic detection and pile hole sonar detection are simulated to validate and demonstrate the effectiveness of the signal analysis method.

The simulations for borehole and pile hole sonic detection utilize a 2D viscoelastic equation (Kohn et al. 2016). The viscoelastic equation is solved using a high-order staggered grid in Cartesian coordinates and the FDTD method. We employ a finite difference scheme with 10th order accuracy in space and 2nd order accuracy in time. To account for wave attenuation and dispersion, we introduce seismic quality factors $Q_{\mathrm{p}}$ and $Q_{\mathrm{s}}$ into the viscoelastic equation using the generalized standard linear solid (GSLS) method (Bohlen 2002). The seismic quality factors are assigned low values at the grid points corresponding to borehole water, pile hole slurry, and infill of karst voids, where notable attenuation and dispersion effects are observed.

The numerical modeling schemes for borehole and pile hole detection are illustrated in Fig. 10.1 To minimize boundary reflections, a perfectly matched layer (PML) is implemented at the computational edges. The numerical models depicted in Fig. 10.1 consist of the borehole, pile hole, bedrock, and karst voids surrounding the borehole or beneath the pile hole. The borehole is filled with water, while the pile hole is filled with slurry. The material parameters ($v_p$, $v_s$, $\rho$, $Q_{\mathrm{p}}$, and $Q_{\mathrm{s}}$) of the bedrock and void infill vary depending on the lithology (refer to Table 10.1 for detailed information). To simulate low- and high-frequency detection, a Ricker wavelet with dominant frequencies of 2 and 40 kHz is transmitted horizontally in the borehole. The recorded horizontal velocity component $v_x$ generates a wave profile (Fig. 10.6a). For pile hole detection, a vertically transmitted Ricker wavelet with a dominant frequency of 3 kHz is recorded in the $v_z$ direction. To ensure simulation stability, different spatial grid intervals, time steps, and PML boundary thicknesses are employed based on the wavelet frequency and velocity (refer to Table 10.2 for specific values).



L. Liu et al., *Detection of Karst Voids at Deep Pile Foundation*,
https://doi.org/10.1007/978-981-97-5834-0_10

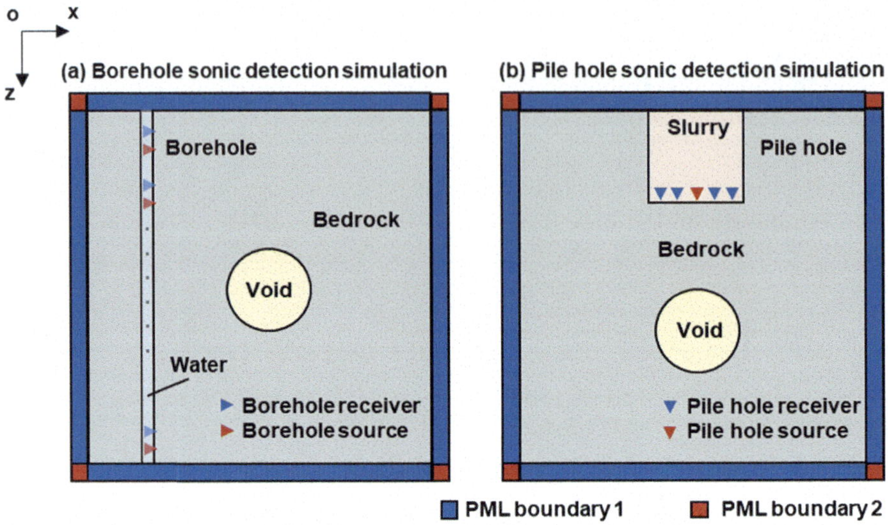


**Fig. 10.1** Scheme of numerical modeling of investigation of karst voids during pile construction. **a** Borehole detection model. **b** Pile hole detection model

**Table 10.1** Parameters of materials in numerical studies

| Scenarios | Material type | P-wave velocity (m/s) | S-wave velocity (m/s) | Density (kg/m$^3$) | Seismic qualities ($Q_p$) | Seismic qualities ($Q_s$) |
|---|---|---|---|---|---|---|
| Model in pile hole analysis | Bedrock | 4000 | 2300 | 2300 | 18 | 18 |
| | Voids infill | 2000 | 1100 | 1200 | 10 | 10 |
| | Slurry in pile hole | 2000 | 0 | 1500 | 6 | 6 |
| Model in borehole data analysis | Refer to Fig. 10.3 (revised based on a realistic field case in Liu et al. 2022) | | | | | |

**Table 10.2** Parameters of numerical simulation for different detection modes

| Detection modes | | Source type | Dominate frequency of source (kHz) | Spatial grid interval (m) | Time step (s) | Thickness of PML (m) |
|---|---|---|---|---|---|---|
| Borehole pile detection | Low frequency | Ricker wavelet | 2 | 0.025 | $2 \times 10^{-6}$ | 0.375 |
| | High frequency | Ricker wavelet | 40 | 0.005 | $2 \times 10^{-7}$ | 0.075 |
| Pile hole detection | | Ricker wavelet | 3 | 0.020 | $2 \times 10^{-6}$ | 0.300 |

The P- and S-wave velocity models (Fig. 10.2) are adjusted based on a previously reported field case (Liu et al. 2022). The revised model reveals four regions with low-velocity anomalies, indicating the presence of multiple karst voids near the borehole. The model has dimensions of 5 m (width) and 20 m (depth). To ensure simulation stability, different time and spatial intervals are employed for low- and high-frequency sonic detection modeling. For low-frequency sonic modeling, a spatial interval of 0.025 m and a time sampling interval of $2 \times 10^{-3}$ ms are used. In contrast, for high-frequency sonic modeling, a spatial interval of 0.005 m and a time sampling interval of $2 \times 10^{-4}$ ms are utilized. Considering the actual configuration of the borehole probe, the distance between the source and receiver is set to 0.2 and 0.1 m for low- and high-frequency modeling, respectively. The density and seismic quality distributions in the model correspond to the velocity variations, with increasing values associated with higher velocities.

Figure 10.3a illustrates the wave profile obtained from high-frequency sonic borehole detection. In the synthetic data, no bandpass frequency filter is applied since frequency noise is minimal in the numerical study. The waves within the borehole dominate the original wave profile, exhibiting multiple vertically decaying events. Due to significant attenuation, the reflection of body waves from the voids is not apparent in Fig. 10.3a. However, the attenuation of waves traveling along the borehole wall (such as direct wave and Stoneley wave) corresponds to the integrity. At depths ranging from 1 to 2, 5 to 7, and 9 to 10 m, both a decrease in amplitude and a delay in the direct wave are observed, indicating the presence of karst voids and fractures. The varying shapes of events at a depth of 16 m suggest a potential low-velocity

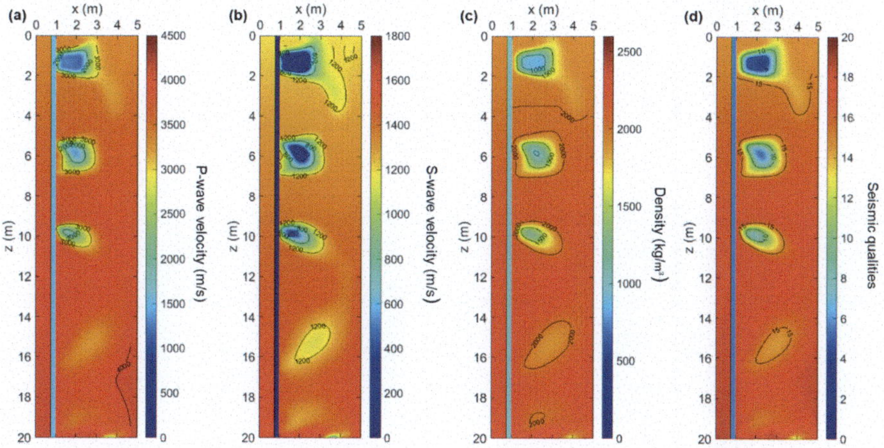


**Fig. 10.2** True velocity model of borehole detection **a** P-wave velocity model, **b** S-wave velocity model, **c** density model, **d** seismic qualities $Q_s$ and $Q_p$

zone near the borehole. To image the integrity of the bedrock surrounding the borehole, signal energy calculations are performed using Eq. (1). Figure 10.3b presents the normalized signal energy results. In general, the signal energy increases with depth as the wave velocity increases (as shown in Fig. 10.2). To mitigate attenuation, larger seismic quality values are assigned in deeper areas. Figure 10.3b indicates the lowest signal energy at depths of 1 and 1.8 m, corresponding to the presence of karst voids filled with low-velocity material. Additional low signal energy traces are observed at depths of 5.5, 6.4, 6.9, and 9.8 m, suggesting fractured rock or locally developed voids. The signal energy below a depth of 10 m is relatively high. The fluctuation at a depth of 16.5 m indicates potential broken bedrock. The analysis of numerical high-frequency sonic borehole data demonstrates that the calculated signal energy results align with the integrity of the bedrock near the borehole wall, offering a more detailed and accurate identification than borehole logs.

Figure 10.4 illustrates the synthetic data obtained from low-frequency sonic borehole detection, along with the corresponding medium filter and F-K filter analysis.

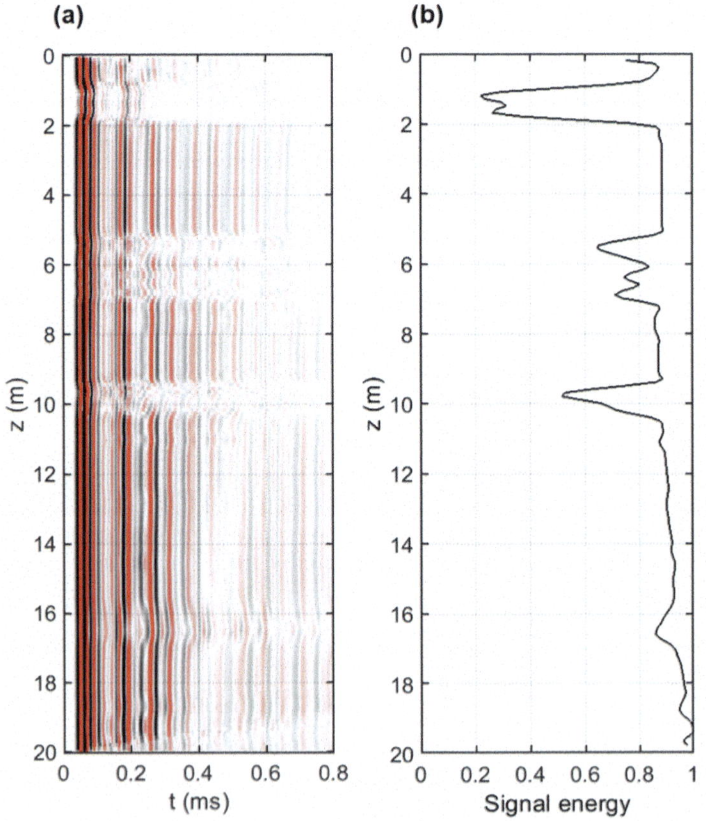


**Fig. 10.3** Signal analysis of borehole detection high-frequency numerical data **a** original profile, **b** signal energy curve with depth

Unlike the high-frequency data, the low-frequency wave profile in Fig. 10.4a does not exhibit significant variations in signal energy with depth. However, the presence of the Stoneley wave traveling along the borehole and reflections from voids outside the borehole can be observed in the original wave profile. To remove the waves within the borehole, a medium filter is applied. The resulting wave profile is separated into two components: the Stoneley wave and void reflections in Fig. 10.4b and multiple waves within the borehole in Fig. 10.4c. During the medium filter, the top and bottom 1 m of the wave profile are removed. The Stoneley wave events can be identified at depths of 1.5, 6.0, 9.8, and 16.5 m, which aligns with the signal energy calculation results from the high-frequency data. The slope of the Stoneley wave events is similar, indicating that the Stoneley wave travels along the borehole with a velocity close to that of the S-wave. Behind the Stoneley wave events, steeper-slope events (approximately 3 ms at depths of 1.5, 3, and 7 and 2 ms at a depth of 17 m) can be observed, representing void reflections. The F-K filter is then applied to the medium-filtered data (Fig. 10.4b) to separate the Stoneley wave and void reflections based on the slope difference, which corresponds to the difference in wave velocities. The extracted void reflections are shown in Fig. 10.4d, where the amplitude of the reflected wave differs between events at different depths. A larger amplitude reflection indicates a more significant impedance contrast (velocity and density) between the void infill and bedrock. Thus, lower-velocity material is inferred to fill the upper voids, consistent with the true model (Fig. 10.2). The signal analysis of the synthetic low-frequency borehole data confirms that the medium and F-K filters can effectively separate wave modes and enhance the visibility of void reflections.

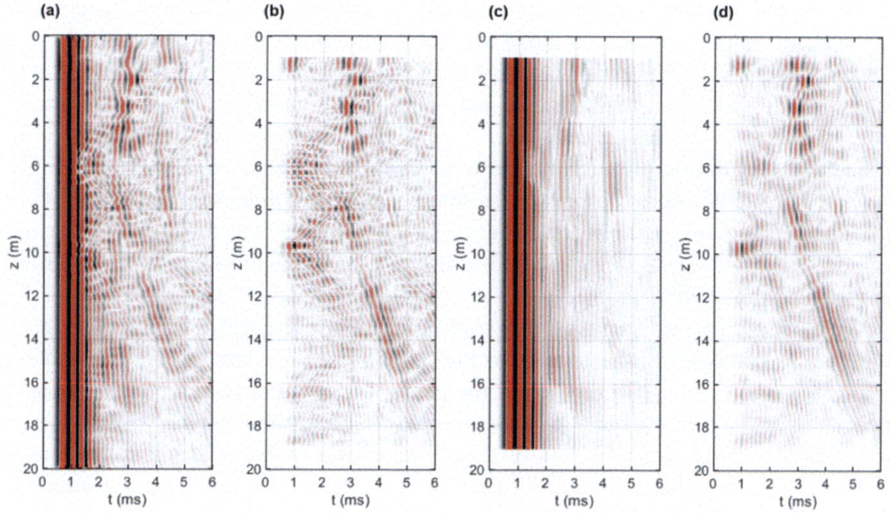


**Fig. 10.4** Signal analysis of borehole detection low-frequency numerical data **a** original profile, **b** median filtered profile, **c** removed wave profile using median filtering, **d** F-K filtered profile

The application of full-waveform inversion (FWI) on low-frequency data aims to provide detailed information about the karst voids near the borehole. The initial models for P-wave and S-wave velocities are displayed in Fig. 10.11a and d, respectively. These initial models are derived by eliminating the low-velocity anomalies from the accurate models depicted in Fig. 10.2. In the FWI process, observed data (as shown in Fig. 10.4a) and synthetic data (generated using the initial model in the first iteration) undergo low-pass filtering before each iteration. A hierarchical inversion approach is employed, where the high-cut frequency of the low-pass filter gradually increases or the inversion process stops when the difference in the objective function misfit between the current and previous iterations falls below 1%. The enveloped objective function-based FWI is initially implemented to improve the quality of the initial model, resulting in intermediate results displayed in Fig. 10.11b and e. Subsequently, the L2 norm FWI is performed based on the updated models to achieve enhanced resolution. Both P-wave and S-wave velocity models are simultaneously updated in each iteration, while density and seismic qualities remain unchanged throughout the inversion process. To maintain stability in the simulation and avoid potential local minima resulting from unexpected updates at the borehole-rock interface, a taper preconditioning operator is employed to control the areas of the model that are updated during inversion. Specifically, the velocity update near the borehole and on the left side of the borehole is muted. After 23 iterations, the normalized misfits of the objective function converge to 0.13. A detailed evaluation of the misfits will be provided in the discussion section.

The resulting inverted P-wave and S-wave velocity models are presented in Fig. 10.11c and f, respectively. Both inverted models exhibit four low-velocity anomalies in the vicinity of the borehole, which are consistent with the true models. Notably, the downward voids at a depth of 10 m can be clearly identified in both the P-wave and S-wave velocity models. The inversion process effectively differentiates the velocity contrast of the void infill, with the upper void being filled with the lowest velocity material. However, it should be noted that the inverted low S-wave velocity anomalies appear larger than the true anomalies. This can be attributed to the inversion process encountering a local minimum when the high-cut frequency is high, indicating a limitation of borehole FWI, as discussed by (Liu et al. 2022). Overall, the numerical FWI study demonstrates the potential of using common-offset borehole sonic data to image the velocity field near the borehole. The inverted P-wave and S-wave velocity models provide valuable information about the shape, size, location, and infill of the voids (Fig. 10.5).

We compare two scenarios in this study to illustrate the signal analysis method: one with a void beneath the bottom of the pile hole (Fig. 10.6a) and one without a void (Fig. 10.6h). The coordinates of the center of the karst void are (2, 3 m), and the diameters of the void and pile hole are 1 m and 1.6 m, respectively. The pile hole is filled with slurry. The material parameters for both scenarios are provided in Table 2. In the sonar wave transmission, the wave is transmitted at coordinates (0, 0) and received at coordinates (0.4 m, 0) and (0.6 m, 0), as shown in Fig. 10.6a and h. We perform one-recording and two-recording analyses on the same model for comparison.

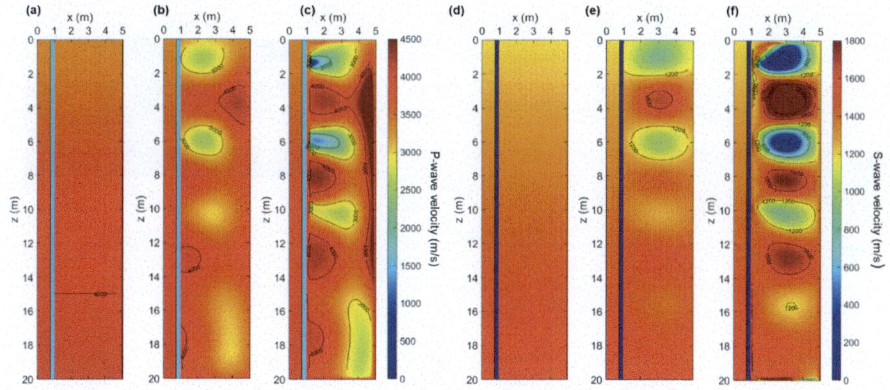

Fig. 10.5 Full-waveform inversion of borehole numerical sonic data **a** initial P-wave velocity model, **b** intermediate inverted P-wave velocity model by envelope-based FWI, **c** final inverted P-wave velocity model by L2-based FWI, **d** initial S-wave velocity model, **e** intermediate inverted S-wave velocity model by envelope-based FWI, **d** final inverted P-wave velocity model by L2-based FWI

The recorded signals are presented in Fig. 10.6b and i for the one-recording analysis. In Fig. 10.6b, the signal contains surface waves propagating at the bottom of the pile hole and the reflection from the void. However, in the absence of a void, the signal in Fig. 10.6i only exhibits surface waves. It is challenging to distinguish the void reflection from the original signal dominated by surface waves (Fig. 10.6b), even when comparing it with Fig. 10.6i. To analyze the time–frequency distribution, we employ the generalized S-transform with a regulatory factor $p$ of 0.5. Figure 10.6c and j shows the results of the S-transform. In Fig. 10.6j, both the frequency and amplitude gradually decay with a consistent time interval when the receiver is positioned at the midpoint between the source and the pile hole wall. However, the presence of a void alters the amplitude and frequency attenuation at points A and B in Fig. 10.6c, which correspond to the top and bottom of the void, respectively. The arrival times at points A and B are approximately 2 and 4 ms, respectively. Based on the velocity of the void infill (2000 m/s) and the time difference between points A and B (2 ms), the estimated size of the void is 2 m. By performing detection in different directions, the location of the voids can be estimated in a similar manner.

The application of the generalized S-transform enhances the visibility of the weak void reflection; however, the arrival time of the reflection remains ambiguous. A more advanced analysis can be performed using two-recording data, as illustrated in Fig. 10.6d–g, k–n. The original signals obtained from the two receivers are presented in Fig. 10.6d and k. The surface waveform recorded by the two receivers exhibits a noticeable phase difference. However, the reflections at 2 and 4 ms are in phase, which can be roughly observed in Fig. 10.6d. The instantaneous phase of the two traces, ranging from $[-\pi, \pi]$, is depicted in Fig. 10.6f and m. Subtracting the two traces directly would result in an instantaneous phase difference range of $[-2\pi, 2\pi]$. The instantaneous phase difference (IIPD) value is calculated as shown in Fig. 10.6g

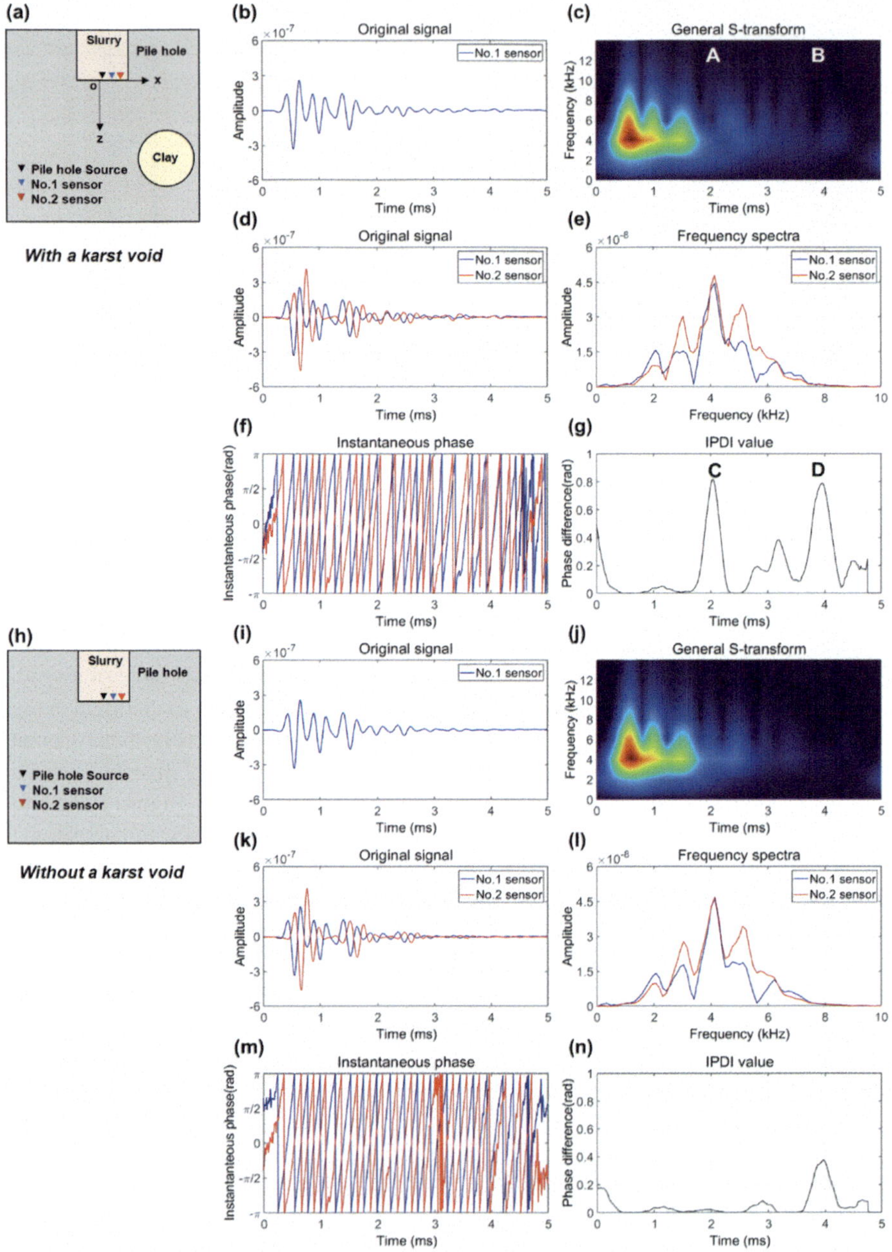


**Fig. 10.6** Pile hole detection signal analysis of numerical data

and n. IIPD can be calculated by normalizing to the maximum and the minimum of IPDI. Peaks C and D in Fig. 10.6g can be identified at 2.002 and 3.994 ms, respectively, while no IIPD peak exceeds 0.4 in Fig. 10.6n. The IIPD analysis of the two scenarios demonstrates that the peaks can quantitatively measure the similarity of the instantaneous phase between the two recordings and identify the P-wave reflection from voids beneath the bottom of the pile hole.

## 10.2 A Complete Field Application

During the construction of a business building using cast-in-place pile foundations in Maguan County, Yunnan Province, Southwest China, a field detection case was conducted in a karst area. The field size, as shown in Fig. 10.7, is characterized by the presence of soft silty clay with angular gravel in the Quaternary alluvial proluvial layer ($Q_4^{al+pl}$), based on regional geological data. The silty clay found in the overburden layer is characterized by a medium dry strength and toughness, appearing as dark gray in color. It is slightly mixed with saturated loose gray breccia, which primarily consists of quartzite and sandy rock. The underlying bedrock is identified as the Cambrian Bocaitian Formation ($\in {}_3b$), characterized by moderately weathered limestone with aphanitic and massive structure.

Figure 10.8 displays three survey lines conducted using electrical resistivity tomography (ERT). The ERT detection utilized the WDJD-3 Multi-function Digital DC Resistivity Meter (Sener et al. 2021) with a 5-m electrode spacing. Resistivity profiles were obtained using the Wenner-β arrays and inverted using a least-squares method with smoothness constraints (Tsourlos 1995). The ERT survey, along with drilling data, indicates that the underlying bedrock at the site consists of soluble

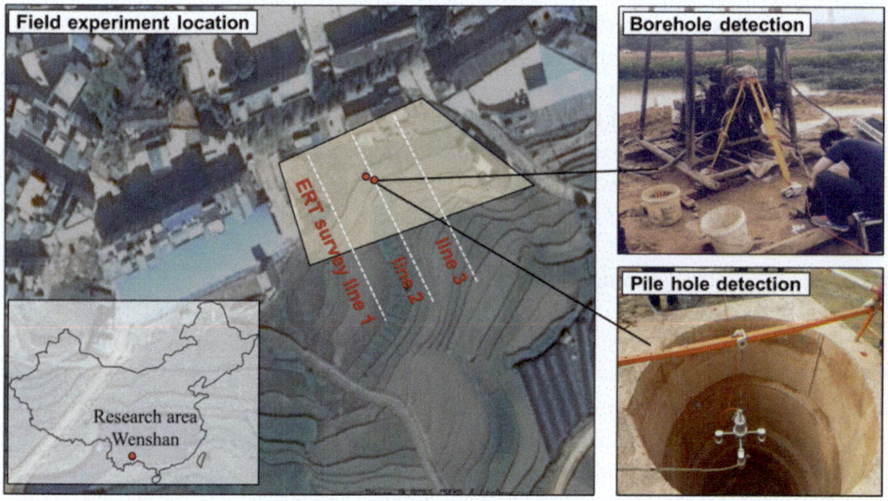


**Fig. 10.7** Location of borehole and pile hole detection field experiment

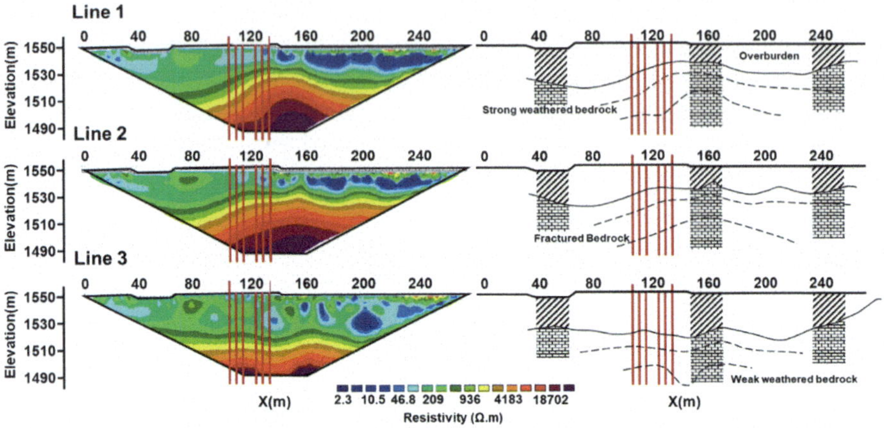


**Fig. 10.8** Geological setting of the field obtained by ERT survey

carbonate limestone. The ERT tomography results reveal a soil-rock interface at a depth of 20 m, with no significant low resistivity anomaly observed (Fig. 10.8). Thus, based on the ERT survey, the bedrock appears to be intact. However, the presence of regional karst voids and fractures is evident from the 18 boreholes (depicted as red lines in Fig. 10.8). The combined length of the rock, as determined from the 18 boreholes, is 622.6 m, with 5.3 m of that length comprising voids. The linear karst rate is calculated to be 0.85%. These findings indicate a weakly developed karst condition at the site, characterized by broken limestone, localized dissolution, and randomly distributed small karst voids. The karst system continues to evolve under the influence of groundwater, which poses risks to the foundation stability and construction. The karst void detection system was employed during both the borehole investigation and pile hole drilling phases, as illustrated in Fig. 10.7.

The borehole detection (Fig. 10.7) took place at the center of the designated cast-in-place pile on the second ERT survey line. During the detection, the high- and low-frequency sources were used, transmitting signals every 0.1 m of depth while raising and lowering the downhole tool. Figure 10.9a displays the high-frequency sonic borehole detection results obtained from depths ranging between 25 and 40 m. Signal energy calculations were performed, and the results are shown in Fig. 10.9b. At depths of 28.0, 29.5, and 37.0 m, low-energy traces indicate localized rock fractures. However, it is worth noting that the normalized signal energy of all traces exceeds 0.3, indicating relatively intact rock at the borehole location.

The original low-frequency detection wave profile is displayed in Fig. 10.10a. To remove the borehole waves, a median filter is applied, resulting in the profile shown in Fig. 10.10c. Stoneley wave events are observed at depths of 27.8, 29.3, and 37.0 m in Fig. 10.10b, which align with the locations of the low-energy traces identified by the high-frequency test. At a depth of 21 m, a void reflection is visible. After applying F-K filtering to the medium-filtered wave profile, the void reflection becomes more distinct (Fig. 10.10d). Consequently, a void can be identified in the vicinity of the

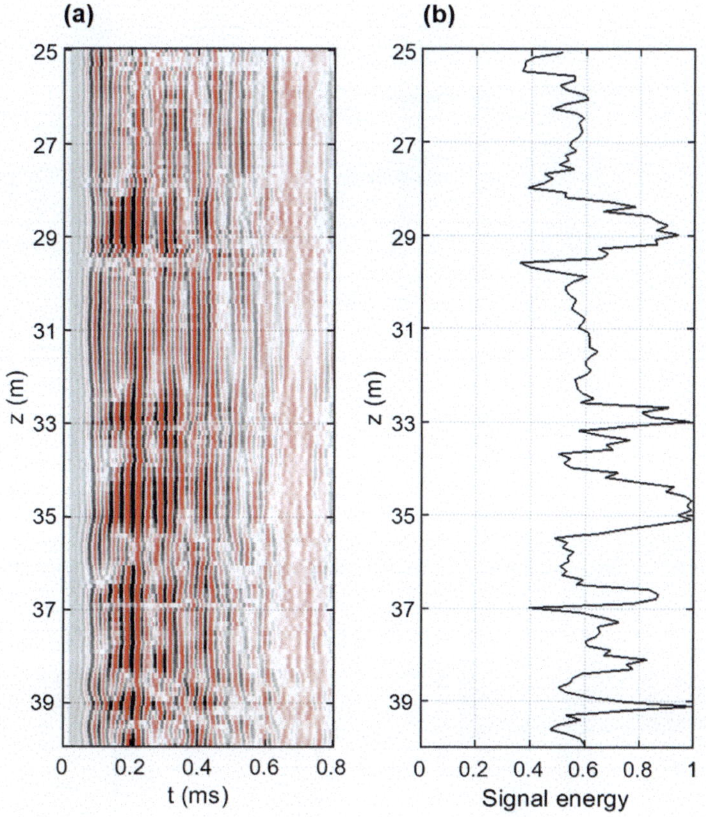


**Fig. 10.9** Borehole detection high-frequency signal analysis of field data **a** original profile, **b** signal energy color bar

borehole. By considering the first arrival of the reflection (approximately 1.2 ms) and the average velocity of the weathered bedrock (2500 m/s), the distance between the near side of the void and the borehole can be estimated to be around 1.5 m.

The full-waveform inversion (FWI) of low-frequency borehole data was conducted using the same 1D linear initial model employed in the numerical study. The borehole data underwent several preprocessing steps. Firstly, the data were resampled to a time interval of 2 × 10–6 s, consistent with the forward modeling. To improve the signal-to-noise ratio, the events preceding the first arrival were muted by zero-padding. A bandpass finite impulse response (FIR) frequency filter with cutoff frequencies of 10 Hz and 4000 Hz was applied. Both the synthetic and field data traces were normalized to the maximum amplitude of each trace. The source wavelet was measured and estimated in a borehole-free water tank, yielding a dominant frequency of 1.7 kHz (Liu et al. 2022). The FWI of the field data utilized a strategy based on successive envelope and L2 norm objective functions, with a high-cut frequency of 0.5 and 1.2 kHz. Unlike the approach employed in the numerical study, a wider taper was implemented around the borehole during the inversion process to mitigate

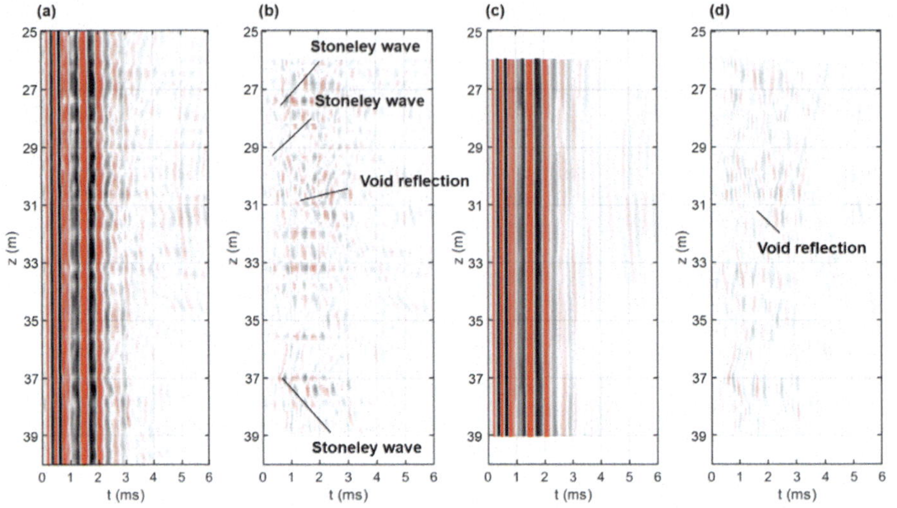


**Fig. 10.10** Borehole detection low-frequency signal analysis of field data **a** original profile, **b** median filtered profile, **c** removed wave profile using median filtering, **d** F-K filtered profile

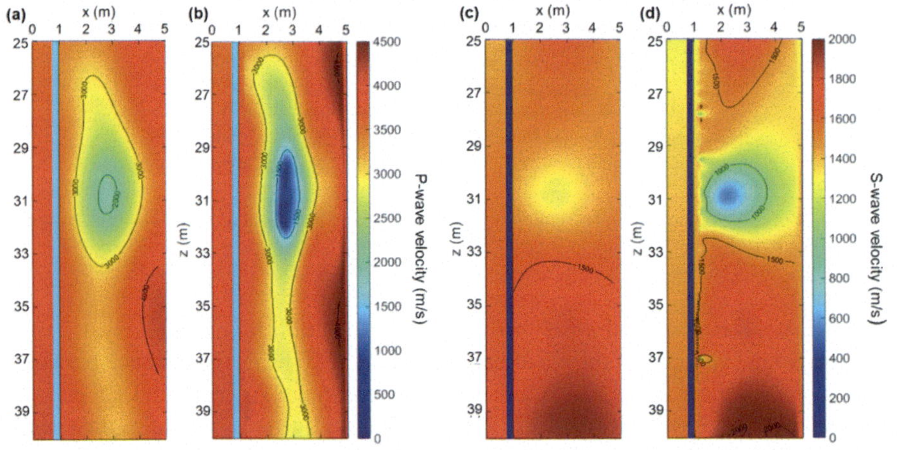


**Fig. 10.11** Full-waveform inversion of borehole field sonic data **a** intermediate inverted P-wave velocity model by envelope-based FWI, **b** final inverted P-wave velocity model by L2-based FWI, **c** intermediate inverted S-wave velocity model by envelope-based FWI, **d** final inverted P-wave velocity model by L2-based FWI

unexpected updates in the immediate vicinity. Additionally, a spatial filtering technique involving a Gaussian function was applied to smooth the updates during each inversion iteration, thereby suppressing artifacts with short wavelengths. After 21 iterations, the normalized misfits of the objective function reached a value of 0.41.

The intermediate and final inverted P-wave velocity models are displayed in Fig. 10.11a and b, respectively. Within the depths of 19 to 22 m, a distinct low anomaly is observed. The contour line of a P-wave velocity of 1500 m/s indicates the presence of a water-filled void with dimensions of approximately 3 m in height and 1.2 m in width. The low anomalies observed both above and below the void suggest fractured bedrock or karst dissolution channels. In Fig. 10.11c and d, the intermediate and final inverted S-wave velocity models are depicted, revealing a low-velocity anomaly within the same depth range of 19 to 22 m. However, the shape of the low S-wave velocity anomaly differs from the P-wave inversion result. The S-wave velocity of the material infilling the void is estimated to be higher than 300 m/s, surpassing the S-wave velocity in water (0 m/s). Two factors contribute to the difference between the inverted P- and S-wave velocities. Firstly, the real field data predominantly contain P-wave components due to the characteristics of the transducers used in the borehole. Although both P-wave and S-wave data are employed in the P-SV modeling and inversion, the reflected P-wave events are more prominent. Secondly, when the voids are filled with water or air (empty), the S-wave velocity can be zero. However, setting the S-wave velocity to zero in the forward modeling and inversion leads to instability. Consequently, inverting for a zero-velocity anomaly becomes challenging.

The comparison of borehole detection results with adjacent borehole logs, borehole optical images, and laser scanning is presented in Fig. 10.12b. Borehole sonic detection is conducted in the right borehole, as indicated. The borehole cores obtained from the detected borehole are displayed in Fig. 10.12c. The rock sample exhibits an intact structure with visible fractures and regional dissolution, thereby confirming the findings of the signal energy calculation from the high-frequency borehole sonic detection. To assess the identification of voids using low-frequency borehole sonic detection, adjacent boreholes are utilized for comparison. Specifically, a borehole located northwest of the detected borehole reveals the presence of a void at depths ranging from 29 to 31 m. The verified borehole, located 2 m away from the detected borehole, is illustrated in Fig. 10.12a. Borehole optical images are employed to examine the infill material within the void. Images captured at the top and bottom of the void are presented in Fig. 10.12d and f, respectively. The presence of water during the borehole optical test aligns with the findings derived from the inversion results. Furthermore, the utilization of borehole 3D laser scanning technology enables the estimation of the void's size and shape, providing a valuable reference for comparing the inverted results.

In the right borehole, a cast-in-place pile with a diameter of 1.6 m was constructed, as depicted in Fig. 10.12b. Pile hole sonar detection was performed using two probes at depths of 26.7 and 35.4 m to examine the presence of voids beneath the bottom of the pile hole during drilling. The test signal acquired from the 1st pile hole sonar probe at a depth of 26.7 m is displayed in Fig. 10.13a. The test signal primarily comprises surface waves propagating at the bottom of the pile hole. Figure 10.13b exhibits the corresponding time–frequency distribution, calculated using the generalized s-transform. The amplitude and frequency attenuation in the time–frequency spectrum is not as periodic as in the numerical signal due to the nonuniform distance between

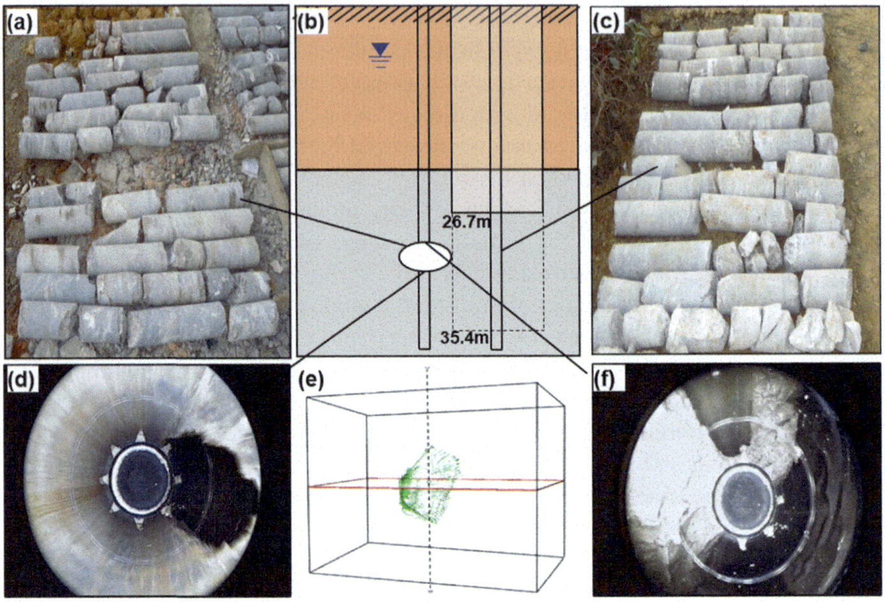


**Fig. 10.12** Scheme of inspection and validation using other methods: **a** borehole for verification. **b** Scheme of detection and verification. **c** Borehole for detection. **d** Optical image at void bottom. **e** Borehole laser scanning results. **f** Optical image at void top

the source and receiver (0.35 m) in relation to half the pile hole radius. However, observations at points A and B (approximately 2.602 and 4.086 ms, respectively) suggest the presence of two reflections originating from the nearby and far sides of voids. To enable more precise identification of the first arrival of the void reflection, a two-receiver detection method was employed. The source-to-first-receiver interval was 0.3 m, and the spacing between the two receivers was 0.19 m. The original test signal is illustrated in Fig. 10.13c, showing in-phase void reflections in the two test signals. Signal reconstruction using EEMD, Hilbert transform, and IIPD calculation were conducted as depicted in Fig. 10.13e, f. The IIPD peaks C (2.534 ms) and D (4.174 ms) quantify the arrival time of the reflections, enabling the estimation of the distance between the pile hole bottom and the void as 3.1 m, based on a P-wave velocity of 2500 m/s. Additionally, the diameter of the voids can be determined as 1.23 m using the P-wave velocity in water (1500 m/s).

The pile hole detection was conducted at a depth of 35.4 m during the pile hole drilling process. The acquisition geometry for this detection was the same as the one performed at a depth of 26.7 m. The two detection signals obtained at 35.4 m (Fig. 10.14a and c) exhibit similarities, making it challenging to determine the integrity of the bedrock. The time–frequency distribution reveals periodic peaks, with some irregular peaks still present in Fig. 10.14b. Two-recording analyses were conducted and are depicted in Fig. 10.14c–f. The IIPD results in Fig. 10.14f show no IIPD peaks exceeding 0.6, indicating that the bedrock beneath the bottom of the pile

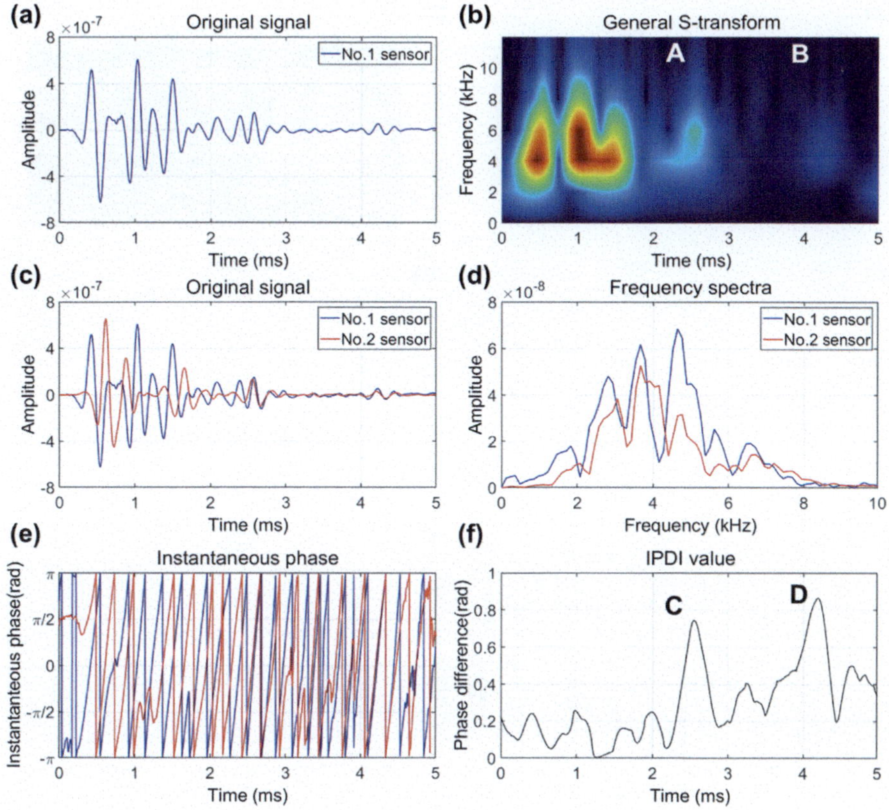


**Fig. 10.13** Pile hole detection signal analysis of field data at 26.7 m

hole is relatively intact. These findings from the two pile hole detections align with the results obtained from borehole sonic detection, borehole core analysis, optical imaging, and laser scanning.

## 10.3 Discussion

Various aspects of the inspection system and analysis method warrant further discussion, as outlined below:

(1) Efficiency of the inspection system and analysis method

Traditional surface and borehole detection methods often suffer from slow analysis due to the complex signal processing and interpretation of geophysical data. In contrast, the inspection system proposed in this study offers improved efficiency through automated probe manipulation. The use of multi-frequency sonic and sonar

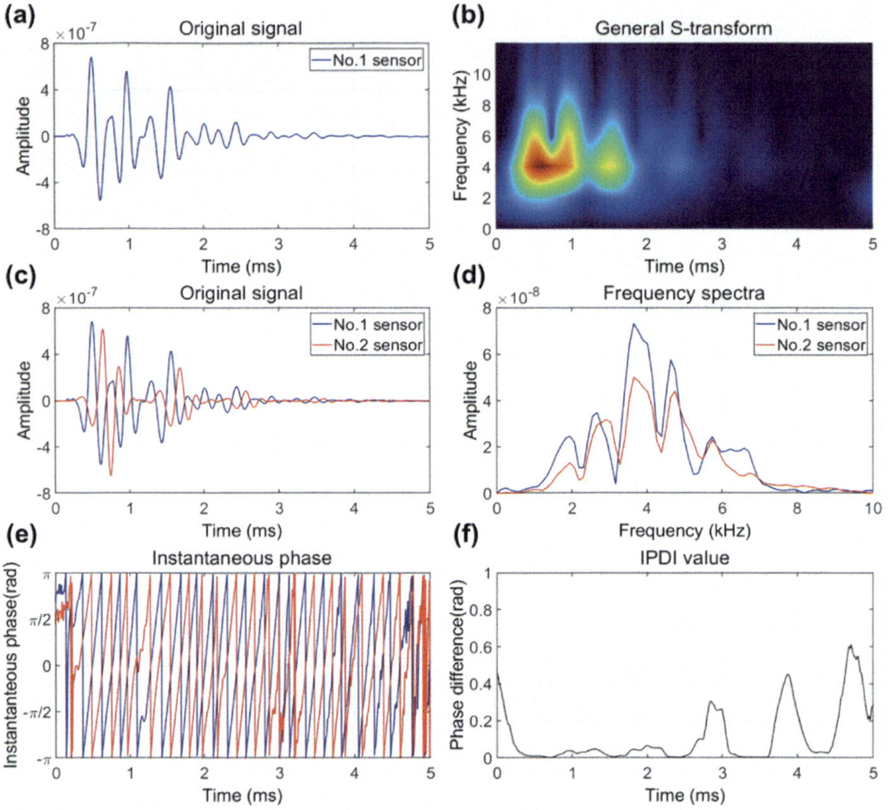


**Fig. 10.14** Pile hole detection signal analysis of field data at 35.4 m

waves, along with their respective recording configurations, enables faster data acquisition. The analysis of detection data is generally rapid, with the exception of the time-consuming full-waveform inversion of low-frequency borehole sonic data. However, other analysis techniques, such as signal energy calculation for high-frequency borehole data, median, and F-K filtering, can be efficiently performed using default filtered velocity thresholds. This enables the identification of voids surrounding the borehole, while the full-waveform inversion provides detailed imaging of void shape and infill. For pile hole sonar detection, the time–frequency distribution and IIPD value can be computed using techniques like the generalized S-transform, EEMD, Hilbert transforms, and the IIPD. Although the identification of void reflection based on the time–frequency distribution may still be ambiguous, the use of the IIPD peaks obtained through the two-recording method offers valuable insights into the arrival of reflections from voids. Overall, the pile hole detection process and corresponding analysis for each pile can be completed within approximately 15 min.

(2) Combination of borehole and pile hole inspection

The inspection process for karst voids during pile foundation construction involves both borehole and pile hole detection, each serving different purposes. Borehole detection primarily aims to gather additional geological information surrounding the borehole, which is crucial for pile design and construction planning. On the other hand, pile hole detection serves as a nondestructive inspection method to assess the quality of the bedrock beneath the pile bottom. The field case presented in this study demonstrates how pile hole detection can validate the results obtained from borehole testing during construction. It is important to note that the detection depth of sonic borehole detection depends on the specific borehole and may be influenced by sedimentary conditions within the borehole. In weakly developed karst areas, boreholes may not be deployed at every pile hole location. However, pile hole drilling is typically conducted for every cast-in-place pile. Therefore, we recommend that pile hole sonar inspections be performed for every pile during or after the pile hole drilling phase. In strongly developed karst areas where borehole investigations are widely utilized, borehole sonic detection is especially recommended due to its effectiveness in detecting karst voids. By implementing both pile hole sonar inspections and borehole sonic detection, a comprehensive assessment of the presence and characteristics of karst voids can be achieved during pile foundation construction.

(3) The weakness of full-waveform inversion

The utilization of full-waveform inversion (FWI) in capturing the velocity model around the borehole allows for the direct identification of voids through the detection of low-velocity anomalies. However, it is important to acknowledge that FWI is a time-consuming and nonautomated process. The complexity of the configuration involves numerous parameters that necessitate manual analysis and expertise. Further research is required to address various challenges and enhance the robustness of the inversion technique.

One significant challenge encountered in FWI is the tendency for the inversion to converge toward a local minimum as the high-cut filter frequency increases. In Fig. 10.15, the misfit curves of FWI for both numerical and field studies are depicted, with annotations indicating different stages of the inversion process based on objective functions and low-pass frequency filters. Notably, the envelope-based and L2-based FWI methods exhibit rapid reduction in misfits within the lower frequency range, while a relatively slower convergence speed is observed in the higher frequency range. Due to limitations in the convergence of ray paths in a common-offset acquisition system, the number of iterations is limited, and only data with frequencies lower than half of the detection data are utilized for inversion, resulting in a decrease in resolution. Therefore, full-waveform inversion for common-offset sonic borehole data remains an ongoing area of research. The presented full-waveform inversion of low-frequency sonic borehole data demonstrates its potential as a valuable tool for imaging karst voids surrounding the borehole in practical applications. However,

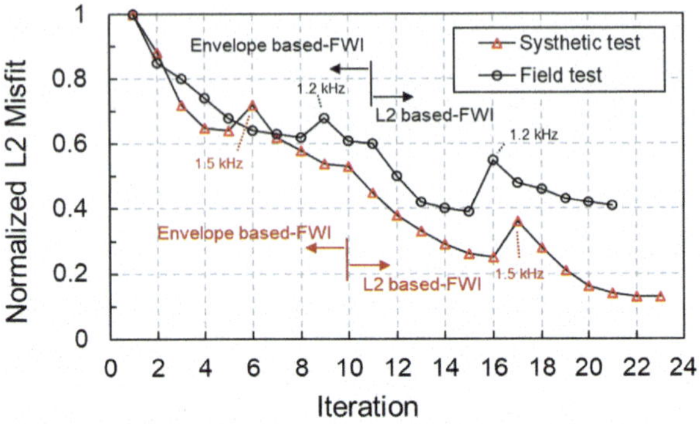


**Fig. 10.15** Normalized L2 misfit curves of inversions in the numerical and field studies

improvements are necessary in various aspects of the inversion process to enhance its effectiveness and applicability in future studies.

(4) Limitations of the current inspection system

The inspection system employed during cast-in-place pile construction offers notable benefits in identifying karst voids. However, it is essential to acknowledge certain limitations associated with the system. Firstly, the current analysis method for high-frequency borehole data is relatively simplistic, relying solely on signal energy. There is a need to extract more valuable information from the waveform that may be concealed. Ongoing research is being conducted on a loop iteration approach that combines high- and low-frequency borehole data to enhance analysis capabilities. Secondly, the determination of the azimuth of karst voids around the borehole can be challenging, particularly in cases where multiple voids are present, using the amplitude difference of reflected waves' first arrival times. To address this issue, an array sonic borehole tool is being designed to mitigate azimuthal ambiguity. However, it is important to consider that long downhole tools may not be suitable for civil engineering applications, even though a sonic receiver array can be incorporated into a lengthy downhole tool. Lastly, uncertainties arise from pile hole sediment and the nonperpendicularity of pile holes, which can impact sonar detection analysis. To address this, a nondestructive test that evaluates the quality of pile holes and the thickness of deposits can assist in inspecting karst voids beneath the pile hole's bottom, thereby reducing uncertainties associated with sediment and nonperpendicularity.

## 10.4 Summary

In this chapter, we present a case study involving the detection of karst voids during the construction of cast-in-place pile foundations. The study was conducted in a karst region and involved the use of an inspection system during both borehole investigation and pile hole drilling processes. The inspection system utilized borehole and pile hole probes to identify a karst void near the pile location. Through the proposed signal analysis method, the location, size, and infill of the karst void were determined. To validate the results, various geophysical techniques such as borehole optical imaging, 3D laser scanning, and borehole logging were employed and compared with the detection findings. The comparison and application of these techniques demonstrate the effectiveness of the inspection system in providing valuable geological information for the design and safe construction of cast-in-place pile foundations in karst areas.

## References

Bohlen T (2002) Parallel 3-D viscoelastic finite difference seismic modelling. Comput Geosci 28(8):887–899

Köhn D, Meier T, Fehr M, Nil D, Auras M (2016) Application of 2D elastic Rayleigh waveform inversion to ultrasonic laboratory and field data. Near Surf Geophys 14:461s-s476

Liu L, Shi Z, Tsoflias GP et al (2022) Detection of karst voids at pile foundation by full-waveform inversion of single borehole sonic data. Soil Dyn Earthq Eng 152:107048

Sener A, Peksen E, Yolcubal I (2021) Application of square array configuration and electrical resistivity tomography for characterization of the recharge area of a karst aquifer: a case study from Meneks¸e karst plateau (Kocaeli, Turkey). J Appl Geophys 195:104474

Tsourlos P (1995) Modeling interpretation and inversion of multielectrode resistivity survey data. Ph.D. Thesis. University of York, U.K

# Conclusion and Further Research Suggestions

## Conclusion

This book focuses on the exploration methods of karst voids at deep pile foundations, aiming to tackle a significant engineering geological issue in the karst area. The study starts from the perspective of survey design and construction features of pile foundations. New exploration observation devices in boreholes and pile holes are developed. Through finite difference numerical simulation, laboratory model tests, field tests, and real project applications, a systematic study was conducted on the detection principles, signal processing methods, and inversion methods based on drilling and pile holes. Finally, a set of methods for detecting karst voids in complex karst areas at pile locations was proposed.

The detection system specifically includes sonic detection of karst voids around boreholes and sonar detection of karst voids beneath the bottom of piles. The borehole detection utilizes existing boreholes at pile locations during the geo-investigation stage to detect karst voids around the borehole, providing a basis for selecting the bearing layer of pile foundations, calculating bearing capacity, and designing construction methods. The pile hole sonar detection utilizes the pile hole sound of the karst voids beneath the bottom of the piles before placing the reinforcing cage in the pile hole, achieving a large-scale inspection of the load-bearing layer at the pile bottom. These two parts of the system enable three-dimensional detection of karst voids on the sides and bottoms of the piles. The following two parts conclusions are obtained:

**Regarding the borehole sonic detection of karst voids:**

(1) This paper proposes a multi-frequency sonic borehole detection system for karst voids around boreholes. It includes high-frequency and low-frequency sonic transducers made of piezoelectric ceramic and magnetostrictive material, directional receiving transducers, depth counters, a 3D electronic compass, and an on-site computer. High-frequency sonic waves accurately identify the karst voids and fractures at the borehole wall. Low-frequency sonic waves detect karst geological anomalies around the boreholes. The wide-band receiving



L. Liu et al., *Detection of Karst Voids at Deep Pile Foundation*,
https://doi.org/10.1007/978-981-97-5834-0

transducers are suitable for both low- and high-frequency detection, and the energy differences between the reflected signals recorded by the four directional receiving transducers can be used to identify the azimuth of anomalies.

(2) The energy of high-frequency sonic detection is strongly correlated with the integrity of the borehole wall rock. When the high-frequency detection signal is strong, the rock is intact. Alternating strong and weak signals within a certain depth range indicate the presence of fractures in the rock. Weak detection signals indicate karst voids filled with sand, clay, or water encountered in the borehole. The color energy bar in the depth direction can visually display the integrity of the borehole wall. The detection accuracy is much higher than that of core sampling. The integrity of the rock obtained from high-frequency sonic detection can help identify the wave components in the low-frequency detection profiles.

(3) Low-frequency sonic detection can identify the karst voids around boreholes by analyzing the reflected body waves from the voids. The low-frequency sonic detection wave field includes direct waves and multiple waves inside the borehole, reflected surface waves generated by the interfaces crossed by the borehole, and void reflection waves outside the borehole. For approximately zero-offset data, the direct waves and multiple waves inside the borehole have similar travel times, and their events are parallel to the borehole, which can be eliminated by median filtering. This enhances the visibility of the reflected surface waves and void reflection waves. Since the interfaces crossed by the borehole generate the reflected surface waves in the borehole, the integrity of the rock obtained from high-frequency sonic detection signals aids in identifying the reflected surface waves. This helps eliminate the interference from borehole surface waves and identify the reflection of void body waves. The receiving transducers facing the void will record higher-energy signals, and this amplitude difference can be used to identify the azimuth of the void.

(4) The multi-frequency sonic borehole detection method was validated in experimental models and a Bridge project in Guangdong, China. The detection results indicate that high-frequency sonic detection can reveal rock fractures not identified in the drilling core. In contrast, low-frequency sonic detection can reveal the voids outside the borehole and their azimuth. Extra drilling was performed to verify the detection results. The borehole detection provides important evidence for selecting bearing layers and construction methods of deep pile foundations in the detailed geo-investigation stage.

(5) A full waveform inversion method of borehole sonic detection combining envelope and cross-correlation objective functions is proposed. The initial model and seismic wavelet are estimated based on drill core data and laboratory experiments. In the multi-stage inversion, envelope-based full waveform inversion is first used to optimize local extrema issues, followed by cross-correlation-based full waveform inversion. The inverted shear wave velocity model can identify karst anomalies around the boreholes. The method's effectiveness is validated through numerical models, where envelope-based full waveform inversion effectively incorporates low-frequency information into the original detection

signals, and cross-correlation-based full waveform inversion further improves the imaging.

(6) The full waveform inversion method was validated by a field test in an expansion project of the Guangqing Expressway. Imaging of karst features around pile locations was achieved using single-hole sonic waveform inversion. A comparison was made between the shear wave velocity profile obtained from full waveform inversion and the longitudinal wave velocity profile from cross-hole seismic tomography. The results confirm the effectiveness of full waveform inversion in imaging single-hole sonic waveform data, accurately reconstructing the shape, depth, and filling conditions of the voids and providing more geological information for selecting load-bearing layers in practice.

**Regarding the sonar detection of karst voids beneath the bottom of piles:**

(1) This book proposes a sonar detection system of karst voids beneath the bottom of cast-in-place piles. It is the first application of sonar technology for detecting voids beneath the bottom of pile foundations. The system includes a high-power, high-frequency underwater sonic source and an underwater transducer. The sonar detection in pile holes effectively utilizes the coupling medium of mud, which has an acoustic impedance closer to the bedrock, as a coupling medium, reducing the impedance difference between different media and improving the sonic coupling rate. This allows the sound waves to smoothly propagate into the bedrock beneath the bottom of the piles, transforming the mud environment's disadvantage into an advantage. The effectiveness of the detection system was verified through engineering case studies.

(2) A 3D numerical simulation model based on the FDTD method was established with the help of field detection data calibration. The numerical simulation reveals that in the sonar detection of voids beneath the bottom of piles, the recorded waveforms by the transducer at the bottom of the pile hole mainly consist of two parts: waves propagating horizontally along the bottom of the pile hole and their multiple reflections on the pile walls, which constitute the main interference in the detection of voids beneath the bottom of the piles. The other part includes waves propagating downward into the bedrock and reflecting on the geological structure surfaces beneath the bottom of the piles, representing effective signals for detecting voids. The depth and size of the voids can be estimated by the propagation time of the reflected waves and the medium's velocity. The time differences of the reflected waves recorded by the receiving transducers at different azimuth angles can be used to calculate the orientation of the voids, enabling comprehensive detection of voids beneath the bottom of the piles.

(3) A generalized S-transform-based signal analysis method for sonar detection was proposed and validated through numerical simulations and field data. When the sonar wave is transmitted at the center of the pile hole, and the source-receiver distance is 0.5 times the pile radius, the multiple reflected surface waves from the pile hole wall exhibit a periodic attenuation of frequency and amplitude equidistant attenuation in the time-frequency spectrum of the generalized S-transform.

However, the arrival of effective reflected waves introduces amplitude enhancement and frequency peaks in the time-frequency spectrum. The corresponding times of these peaks in the time-frequency spectrum can be used to determine the arrival times of the reflected waves, thereby determining the depth of the karst voids beneath the bottom of the piles.

(4) A two-sensors observation system for sonar detection beneath the bottom of piles was established, incorporating a dual-channel receiving transducer with a spacing equal to half the wavelength of the surface wave. This ensures that the instantaneous phase difference between the two receiving transducers for the reflected P-waves from the void beneath the bottom of the pile is close to zero. In contrast, the instantaneous phase difference for the background multiple reflection surface waves is $\pm\pi$. Based on this characteristic, an Instantaneous Phase Difference Intensity (IPDI) index was defined to quantitatively characterize the similarity of the instantaneous phase of the two sensor recordings. A higher IPDI value indicates a higher similarity in the instantaneous phase of the two recordings within a certain period, indicating the arrival of the reflected waves from the void beneath the bottom of the pile. Weak reflections from the top and bottom of the void beneath the bottom of the pile can be automatically identified using this method.

(5) The IPDI method can indicate the reflected waves from voids beneath the bottom of the piles even in severe surface wave interference and other detection noise. The depth, horizontal position, and size of the voids can be calculated using the IPDI method. Parameters such as pile diameter, mud properties in the pile hole, the void infill, source-receiver distance, and distance between the two receivers significantly impact the detection signals. Sensitivity analysis of the parameters shows good consistency with most parameter variations. The key to the IPDI method lies in the distance between the two receivers. When the distance is set exactly to half the wavelength of the surface wave, the IPDI peak is most pronounced. When this distance is within the range of 1/3 to 2/3 of the wavelength, the effective peak value of IPDI is more than twice as high as the background noise.

## Further Research Suggestion

This book proposes detection methods for voids at deep pile foundations in complex karst areas, including sonic detection around boreholes and sonar detection beneath the bottom of piles. The effectiveness of these methods has been validated through numerical simulations, laboratory experiments, field tests, and engineering applications. However, due to the complexity of karst development and detection, further research is still needed in the following aspects:

**Regarding the borehole sonic detection of karst voids:**

(1) The analysis of high-frequency signals in the detection around boreholes is relatively simple, considering only the signal energy without a deep investigation of the waveform. Further potential in signal processing can be explored to achieve better imaging results for geological formations near the boreholes.
(2) The numerical simulation of borehole sonic detection in this paper is based on finite difference in a 2D Cartesian coordinate system. Future research can consider three-dimensional numerical simulations in a cylindrical coordinate system to achieve better simulation accuracy for borehole detection. Additionally, using variable grid-based calculations can help reduce computational costs. Effective forward modeling methods can extend to full waveform inversion as the forward operators.
(3) In the full waveform inversion of borehole sonic data, the inversion results and accuracy of P-wave velocity are unstable, mainly due to the longer wavelength of P-waves. Furthermore, current full waveform inversion methods still encounter issues of local minimum, especially in the relatively high-frequency range, which has a greater impact on the inversion results of P-waves. The number of iterations is also limited and requires further research to propose better inversion strategies.
(4) The accuracy of inverting near-borehole structures significantly impacts the inversion of the far geological structure around the borehole. A potential for future research is how to use high-frequency detection data to provide velocity models near the borehole and feed them back to the full waveform inversion of low-frequency data. A frequency-space iterative inversion approach can be considered if a full waveform inversion of high-frequency borehole data can be achieved.

**Regarding the sonar detection of karst voids beneath the bottom of piles:**

(1) The generalized S-transform time-frequency analysis for void detection beneath the bottom of piles has strict requirements for observation geometry. Reasonable results can be obtained only when the source-receiver distance is 0.5 times the pile radius. However, the analysis is still prone to be affected by the multiple reflection surface waves at the bottom of the pile hole. The information provided by single-channel signals is limited, and future research should focus on array-based detection at the bottom of pile holes.
(2) The current research indicates a relationship between the calculation stability of IPDI and the boundary effects of EEMD, which requires further investigation. Additionally, the current research on void detection beneath the bottom of piles has not considered unfavorable conditions that may occur in practical engineering, such as sedimentation beneath the bottom of pile holes, irregular shape of pile holes, or inclination of pile holes. Future research should consider more unfavorable scenarios to enhance the adaptability and reliability of the methods.
(3) Due to various factors such as the progress of equipment and methodology research and actual engineering projects, the engineering cases presented in this

paper for borehole detection and pile hole detection are from different stages of different projects, lacking a complete demonstrative case. Future research aims to track the entire process in a large-scale project, from survey design to construction verification.

## Acknowledgment

First, the authors would like to gratefully acknowledge the support of the National Natural Science Foundation of China (Grant No. 42172296 and 42207211).

The manufacturer's authorised representative in the EU is Springer Nature Customer Service Centre GmbH, Europaplatz 3, 69115 Heidelberg, Germany. If you have any concerns regarding our products, please contact ProductSafety@springernature.com

Printed and bound by CPI Group (UK) Ltd, Croydon, CR0 4YY
01/10/2026
02225926-0003